Springer Tracts in Civil Engineering

Springer Tracts in Civil Engineering (STCE) publishes the latest developments in Civil Engineering - quickly, informally and in top quality. The series scope includes monographs, professional books, graduate textbooks and edited volumes, as well as outstanding PhD theses. Its goal is to cover all the main branches of civil engineering, both theoretical and applied, including:

- Construction and Structural Mechanics
- Building Materials
- Concrete, Steel and Timber Structures
- Geotechnical Engineering
- Earthquake Engineering
- Coastal Engineering; Ocean and Offshore Engineering
- Hydraulics, Hydrology and Water Resources Engineering
- Environmental Engineering and Sustainability
- Structural Health and Monitoring
- Surveying and Geographical Information Systems
- Heating, Ventilation and Air Conditioning (HVAC)
- Transportation and Traffic
- Risk Analysis
- Safety and Security

Indexed by Scopus

To submit a proposal or request further information, please contact:

Pierpaolo Riva at Pierpaolo.Riva@springer.com (Europe and Americas)
Wayne Hu at wayne.hu@springer.com (China)

Ivan Pistone

Urban Coasts in Socio-ecological Transition

A Methodological Framework to Support Planning and Management of the City-Sea Interface

Springer

Ivan Pistone
DiARC—Department of Architecture
University of Naples Federico II
Naples, Italy

ISSN 2366-259X ISSN 2366-2603 (electronic)
Springer Tracts in Civil Engineering
ISBN 978-3-031-70782-7 ISBN 978-3-031-70783-4 (eBook)
https://doi.org/10.1007/978-3-031-70783-4

This Springer imprint is published by the registered company Springer Nature Switzerland AG
The registered company address is: Gewerbestrasse 11, 6330 Cham, Switzerland

Foreword

The coastal landscape constitutes a research topic for numerous disciplines: from biology to economics, from sociology and geography to design disciplines such as architecture and urban planning. Literature and the visual arts, from painting to photography and cinema, are also constantly concerned with it, highlighting values, opportunities, but also risks, challenges and, above all, the fascination of a boundary between land and sea, between two ecosystems that man has sought to understand since his earliest past, overcoming the fears of an unknown world.

The coast is first and foremost a limit, a threshold, an edge that mankind attempts to understand and that constitutes one of the most interesting cultural phenomena analysed and studied by the territorial disciplines. The narrative of coastal landscapes is developed in art as an interweaving of representations, between leisure and despair, and conveys aesthetic and iconic approaches that affect our imagination. The arts often depict the sea as 'the beast', to quote the novel *Moby Dick*, or as a savage force, as in *The Wanderer on the Sea of Fog* by Caspar David Friedrich or *The Scream* by Edvard Munch. In literature as in painting, the coastline becomes a metaphor for a limit, an expression of the confrontation with the infinite, unknown and boundless domain of the sea.

There are other ways of describing and studying this limit, which do not instinctively respond to our imagination but scientifically and rationally address the understanding of the coastal landscape, and this is what the subject of this book explores. The approach is fully scientific with reference mainly to the disciplines of geography and urban planning: two forms of knowledge that aim to understand and envisage transformations of the coastal territory in order to respond to numerous social demands.

In its opening, the volume appropriately defines the complexity of the coastal landscape, highlighting its multidimensionality and dynamism between competing uses and territorial risks, the latter exacerbated by the current climate change, complicating the planning and management of efficient littoral changes.

Coastal and terrestrial ecosystems have undergone considerable human exploitation in recent centuries, and especially in the last seventy years, man-induced modifications of marine and littoral ecosystems have increased. Beaches, in particular, have become attractive and crowded places, where the demands for technology and energy are increasingly pushing and affecting changes. Ports and the fishing industry compete globally, wind farms dominate the renewable energy market and pollution invades the oceans. Consequently, the problems of the coastal landscape touch contemporary issues such as climate change, sea level rise, urbanisation, coastal management, environmental protection, renewable energies, green and blue infrastructure, the role of ecosystem services, sustainable mobility and heritage identity along the shore.

In this multifaceted context characterised by multiple needs, classic planning tools like town plans and infrastructure projects often reveal their inability to provide adequate answers to the demands of the future. What is needed instead is a combination of bottom-up and top-down processes that can build a shared vision among stakeholders in order to establish common goals. This volume deals with the definition of a survey and research methodology for shared solutions, combining geospatial analysis and participatory techniques, as can be clearly seen in the chapters that structure the table of contents.

With this book, the author intends to provide a methodology for analysing coastal areas by defining an approach that combines understanding of existing legislative and planning frameworks, which regulate coastal transformations, with the analytical knowledge of geographic sciences, using modern and innovative digital technologies, and the participatory and co-design approaches with local communities. The analytical approach is always aimed at transformation with a proactive attitude towards landscape design.

The sustainable design and management of coastal landscapes is the scientific tool to pursue the main objective of coastal planning: the psychological, social, physical and economic well-being of coastal communities.

Through the clear definition of its research question and its coherent analytical-descriptive structure, including the application to the case studies of Naples and Marseille, this volume succeeds in providing a methodology that can be replicated in other contexts and that can involve scholars, citizens and experts in the co-design process for the proposition of shared physical design solutions. The proposed framework can and should be understood as a scientific *research for design* tool that uses multidisciplinary knowledge and digital means to assess the impacts of coastal landscape transformation projects and programmes and to propose integrated solutions to improve safety, biodiversity and recreation and tourism opportunities so that coastal communities can thrive in a well-balanced way.

Antonio Acierno
DiARC—Department of Architecture
University of Naples Federico II
Naples, Italy

Preface

Contemporary urban coastal areas are increasingly becoming crossroads of complex social and environmental phenomena, at the same time influenced by economic-productive demands. These spaces are characterised by a high population density, which fuels the expectations of local communities with regard to coastal development, due to their strong attractiveness for users, as well as their great landscape and functional value. The spatial and morphological characteristics of urban coasts have generated interpenetrating and increasingly complex land (and sea) uses, leading to the most disparate spatial and functional configurations: urban beaches, port areas, quays, sea promenades, but also wastelands and semi-natural spaces compose a mosaic that is as articulate as it is interesting from an urban design perspective, despite the fact that the strong functional competition can drastically reduce the possibilities of fair accessibility to the shores, compromising their identity value. From an ecological-environmental point of view, urban coasts are particularly vulnerable to the worsening of climate change, which inevitably compromises the feasibility of urban interventions, reducing the space available for human activities along city shorelines. On these premises, the volume aims at deepening the knowledge of planning in coastal areas, proposing the concept of the city-sea interface as a key for interpreting the interactions between land and sea, with a view to a responsible governance of the territory in its areas most connected to urban water dynamics.

The innovation of the work lies within the framework of coastal planning studies from both the land and maritime sides. Indeed, the theoretical efforts seek to define a coherent link between land planning, rooted in traditional and contemporary urban studies, and the more recent maritime spatial planning, which cannot be considered separate from the former. In fact, the exploration of the city-sea interface theory aims for this objective, bringing the territorial perspective back to a more human and local scale. Thanks to its flexibility, this theoretical construct, as an urban amphibious, can be considered a platform of socio-ecological connection within which users' demands can be truly integrated in a careful and conscious planning, aimed at restoring high-quality coastal public spaces.

In this regard, a further new element that the book introduces is a methodological framework designed to systemise the normative, geo-spatial and psycho-social criticalities inevitably connected to coastal planning, analysing the main morphological and functional characteristics of the urban amphibious, in order to understand the specific relevance of each existing use, in relation to the needs of coastal society but also to the impacts that exogenous factors (such as risks coming from the sea or related to climate change) may have on the usability of urban coastal areas and on their aesthetic, landscape and social value. Its articulation suggests the possibility of integrating it with planning and decision-making processes, constituting a solid basis to support coastal-sea spatial planning in downscaling to the local and social dimension of the city-sea interface.

The opportunity to explore these issues arose through the doctoral research of the author, carried out under a co-tutelle agreement, intertwining the academic course of Urban Planning at the University of Naples Federico II and the National Research Council (CNR), on the Italian side, with the course of study in Geography at the Aix-Marseille Université. Moreover, the work here presented is connected to the participation in the international research call Galileo 2021, a scientific cooperation programme between Italy and France organised by the Université Franco-Italienne. The project, entitled 'Re-SEA-ourcing CITY. City-sea interface as a resource for people: urban regeneration in the context of ecological transition', was named the winner of the call for proposals and involved the collaboration of the Department of Architecture of the University of Naples Federico II and the CNR), on the Italian side, with the Centre National de la Recherche Scientifique (CNRS) and the Aix-Marseille Université, on the French side. The research deepened the theme of the city-sea interface as a resource for the regeneration of urban coasts, characterised by community and economic interests, mitigating coastal risks in the context of the ecological transition and reinterpreting the coasts of large cities in a socio-recreational and landscape-environmental key for sustainable management and efficient planning: the aim was to develop socio-ecological analyses on the city-sea interfaces of Marseille and Naples, studying social needs and physical-functional features of vulnerable urban coastal territories, through a strongly multidisciplinary research team, composed of urban planners, geographers, environmental and community psychologists, architects and cartographic engineers.

This volume is addressed to policy makers who are increasingly called upon to operate in contexts of both environmental and social criticality along urban coastlines. In fact, the development of urban settlements along the coast cannot ignore community dynamics and the growing social demand for spaces dedicated to recreational uses. The quality of the coastal areas affects the well-being of the numerous users who daily frequent the city spaces along the sea, for the most diverse reasons. In this sense, a key role is played by the understanding of the city-sea interface as an area in which environmental risks from the sea tend to have the greatest impact, because of the numerous interests and the conflicts within society that decision-makers must efficiently resolve.

At the same time, the book is also intended for urban planners and designers who are facing the complexity of coastal land projects and are developing maritime spatial planning tools that consider current criticalities. The definition of a methodological framework for the legal, geographic and psycho-social analysis of this area of the urban coast focuses on the social and recreational component towards which the research question is directed; however, its flexibility opens up possible scenarios of integration with further aspects of maritime-coastal planning, in relation to commercial, economic, productive and logistic development, always respecting the socio-environmental criteria of the coastal context. If, therefore, from an operational point of view, this approach constitutes a possible scaffolding to build an integrated coastal-maritime planning that is attentive to both its spatial-functional components, from a theoretical point of view it aims to fit within the current multidisciplinary research field aimed at combining the management of coastal spaces oriented towards social development and environmental protection with the potential of maritime spatial planning.

Naples, Italy Ivan Pistone

Acknowledgements

This book is the result of extensive studies in complementary disciplinary fields related to the complex coastal issue and it has received important contributions that have strengthened its theoretical and methodological foundations.

Mention must therefore be made to the input of the research group of the Galileo 2021 programme, composed of the Italian researchers Antonio Acierno and Alessandra Pagliano for the Department of Architecture of the University of Naples Federico II and Massimo Clemente and Gaia Daldanise for the Institute for Research on Innovation and Services for Development (IRISS) of the CNR, and the French researchers Samuel Robert, Alexandra Schleyer-Lindenmann, Marie-Laure Apers-Trémélo and Brieuc Cabioch for the ESPACE laboratory (UMR 7300) of the CNRS and the Aix-Marseille Université. In particular, the Italian group is credited with the support in the socio-environmental and regulatory-planning study of coastal space, while the expertise of the French group can be read in the psycho-social and geo-spatial analysis of urban littorals.

In addition, it is well worth mentioning the role of Antonio Acierno and Samuel Robert, as co-tutors of the doctoral work underlying this volume: the former has been a guiding figure in the application of the urban planning perspective to the topic of designing and managing coastal public space, with genuine passion and attention; the latter has instead encouraged a geographic vision towards the topic of the city-sea interface in an enthusiastic and meticulous manner, opening up the work to broader and more complex research scenarios.

Contents

Chapter 1
Introduction. The Contemporary Urban Coast

Abstract Coastal urban areas are crucial for urban planning, as they represent spaces where new ways of urban transformation can be implemented with the participation of the community. They are, in fact, densely populated spaces influenced by social, environmental and economic-productive dynamics: in addition, their landscape and functional value stimulates expectations of coastal development. They offer opportunities for ecological transition through environmental restoration strategies, the use of renewable energy and water quality improvement programmes. Coastal planning and management are thus closely intertwined with the evolutionary processes of the shore, requiring the integration of interdisciplinary approaches in contemporary littoral design.

Keywords Urban coasts · Coastal planning · Coastal management · Socio-environmental demands · Interdisciplinary design

1.1 Urban Coasts as Sustainable Transformative Opportunities for Urban Well-Being

Contemporary urban coastal areas increasingly convey complex social and environmental features, at the same time influenced by economic and productive stakes. These spaces are characterised by a high population density that raises expectations of the local community towards coastal development, due to its high appeal for users, as well as its great landscape and functional value (Small and Nicholls 2003).

The spatial and morphological characteristics of urban shores have generated intertwined and increasingly complex land (and sea) uses, leading to the most disparate functional configurations: urban beaches, port areas, piers, sea promenades, but also wastelands and semi-natural spaces compose an articulated yet interesting urban mosaic, even though the strong functional competition can drastically reduce the possibilities of equal accessibility to the coasts, compromising their identity value. From an ecological-environmental point of view, urban coastlines are particularly

I. Pistone, *Urban Coasts in Socio-ecological Transition*,
https://doi.org/10.1007/978-3-031-70783-4_1

vulnerable to worsening climate change, which inexorably complicates the feasibility of plans and interventions, reducing the space available for human activities along the shore (Giannakidou et al. 2019). The massive densification of the urban fabric can damage the littoral environment, constituting by its very conformation an element of disconnection of the coastal ecological networks (Pineschi 2013). To this sum the presence of defence works against hydraulic risk that are not always adequate to protect the urban-coastal ecosystem (Pilkey and Cooper 2014). These critical issues tend to increase much more rapidly in large cities, due to the multiple and broader interests involved.

The study of urban coastal areas is a well-known research topic, especially from the perspective of sustainability, urban regeneration, recovery of brownfields, port and harbour areas and the evolution of the urban waterfront in general. In any case, city shorelines are characterised by a very high landscape value, consistent with an intrinsic social and recreational character that can still inspire differentiated uses. Its ecological role is fundamental, especially since most of the surrounding areas are entirely urbanised. Thanks to marine biodiversity, coasts can provide a privileged contact point with nature for the inhabitants of large coastal cities (Edward 2017). Moreover, they are places where the ecological transition of the city can be experienced through environmental restoration strategies, the use of renewable energy from the sea or programmes to improve the quality of bathing water. In this context, urban coastal areas represent a unique opportunity for the community in terms of urban planning. They are spaces where a new type of urban transformation can be implemented together with citizens. From this perspective, urban planning and coastal management are closely interconnected within contemporary coastal evolutionary processes: actions included within coastal programmes must be increasingly enriched by techniques, methods and approaches from other disciplines, in relationship with the transformation of the urban coastline, transcending the limits of scale and including communities in the contemporary coastal project (Kay and Alder 2017). This seems particularly interesting, as land-based planning is rooted within the traditional urban planning theory, despite the fact that there are still gaps within which planners and decision-makers must intervene; on the contrary, the management of the maritime space is a field yet to be explored, an opportunity that should be integrated with the urban littoral project, from a social, cultural, functional and environmental point of view. Planning the city's stretch of water may indeed provide significant insights in the management of competing coastal uses. With regard to these reflections, the book introduces the concept of *city-sea interface*, that is the coastal connection area between the land and the sea inside dense urban agglomerations, in which the relations between the two systems appear to be particularly intense. The importance of this concept can be read in relation to the search for a point of contact, for a *trait d'union* between the consolidated terrestrial planning and the emerging maritime spatial planning (Fig. 1.1).

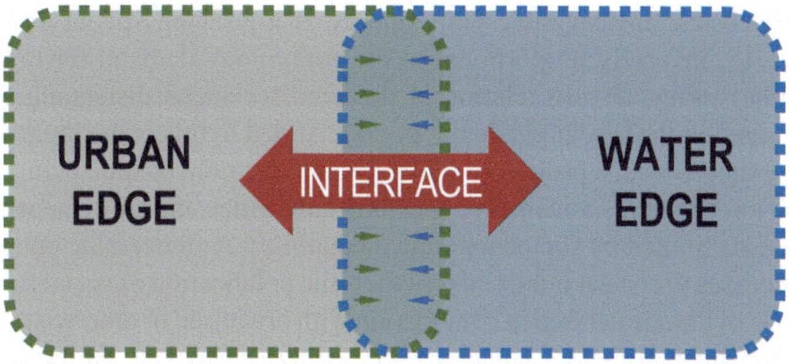

Fig. 1.1 The interactions between the water system and the urban system influence the city-sea interface, which in turn reciprocally influences both systems (elaboration of the author)

1.2 The New Coastal Question, Socio-Environmental Crisis for an Ever-Changing Society

Over the last century, interventions for coastal transformation have mainly concerned the modelling of urban waterfronts, the enhancement of logistical and economic needs, and the recovery of coastal aesthetic value. However, the current condition of the urban littoral brings out the need to focus planning actions more on the theme of coastal social needs.

Between 1970s and the beginning of 1990s, the development of urban coastal areas was oriented towards tourism, recreation, but also economic progress. Numerous investments were promoted in favour of the renovation of harbours and coastal areas, realising spaces for entertainment, reusing existing areas, especially former industrial sites, and implementing commercial and socio-cultural buildings; this led to significant economic growth and the enhancement of the visual quality of coastal areas (Porfyriou and Sepe 2017). Between the end of the twentieth century and the dawn of the new millennium, a redirection of planning efforts can be seen, in order to balance the results of the previous project trend in urban coastal areas: if initially the economic profit and aesthetic improvement were pursued, studies and projects for the development of more accessible and inclusive public spaces were subsequently produced; there was a growing interest in social issues, even if the main objective was limiting gentrification (Morena 2011). On the other hand, the last decade is characterised by an ideal break from the past, as social demands tend to become more and more complex: the economic component of the coasts contrasts with the evolving needs of citizens, also altered by the effects of climate change, distorting the modalities of social interaction and influencing the perception of spaces and attachment to place of the users (McElduff and Ritchie 2018).

The community evolves in relation to littoral changes. Increasingly, the urban coastal society is involved in wide-ranging issues, such as environmental and ecological risks; to these are added the criticalities related to the consumption of land that today is a rare but necessary resource to ensure a fair use of leisure and social places along urban coasts; an inevitable consequence is a growing spatial and environmental injustice inversely proportional to the community demand for high quality coastal public spaces, also in relation to the need for social distancing for health reason, emerged in the recent pandemic period. Added to this is the issue of limited accessibility that worsen a progressive social fragmentation: despite being a right of citizens and a relevant responsibility for public authorities, the enjoyment of coastal benefits appears limited by socioeconomic inequalities; moreover, the uneven distribution of accesses to coastal public spaces fuels the polarisation of social functions in limited points of the coast, often in conjunction with privatised or otherwise unequally usable areas (Kim et al. 2019). The issue of equalitarian access to public places such as urban beaches falls within the field of environmental justice, since the enjoyment of coastal public space, as a common asset of society, turns out to be an important civil right for coastal users. Moreover, guaranteeing the availability of such urban areas is one of the duties of local government, as they have great influence on the physical and psychological well-being of citizens, regardless of their social status (Kim 2015) (Fig. 1.2).

The urban coast must therefore assume features of adaptability and flexibility to satisfy changing community needs, facing external impacts in an effective way. This is particularly significant in relation to how the new post-Covid coastal community experiences sociality in its everyday life, absorbing the social impact of the evolved condition of urban life through new ways of using littoral public places, as well as the

Fig. 1.2 Thousands of users enjoy the urban sea along the Spanish beach of La Barceloneta, in Barcelona. It is an important coastal tourist-recreational area. However the constant overcrowding undermines the fair distribution of public spaces within the local community (ph. M. Sokolov, from Wikimedia Commons, licensed under CC BY-SA 4.0)

ecosystem resources they can provide: it is important for urban planners to analyse in detail the spatial composition of the city-sea interface with particular attention to the places of sociality that could become a new city network at the service of coastal society (Fig. 1.3).

Fig. 1.3 The Palm Jumeirah area in Dubai is one of the most famous example of coastal transformation towards the sea: the creation of a completely artificial island had significant economic implications but it also raised numerous landscape and social questions (from Wikimedia Commons, licensed under CC BY-SA 3.0)

1.3 The Euro-Mediterranean Context

The particular geographical area analysed by the work is the Mediterranean basin, with specific focus on the European urban coastline. For the inhabitants of the cities that rise in this zone, water areas represent spaces that are essential to their existence, economically, but also culturally and socially: it is no coincidence that, for centuries, human settlements on three continents have been concentrated along the coastlines facing this sea, due to the richness of resources and advantageous position, expanding the historical and cultural heritage that characterises this specific zone, but also the strong anthropic and environmental pressure. In fact, it is along these shores that most Europeans live today and the trend is increasing: the presence of important cultural, landscape and economic values has led to the occupation of a consistent part of the available coastal land by urban centres and infrastructures of various kinds.

The Mediterranean sea is directed towards a rapid development of new economic activities. The European Union's Blue Growth strategy[1] foresees further development of sectors that have the potential to create sustainable jobs and generate positive evolution, related to renewable energy sources, aquaculture, maritime-coastal tourism and more. However, the development of such activities implies an increasing competition for space with currently existing traditional marine uses, such as shipping or fishing, which already occupy large portions of sea. Hence, it emerges an interest in deepening the planning studies of these areas. However, it has to be noted that actual protective policies and plans for Mediterranean coastal areas started a relatively short time ago, since the attention of institution was previously focused mainly on inland areas, where pressures related to tourism or other similar functions are not that strong. Today about 80% of European areas bordering the Mediterranean basin is degraded in ecological sense and compromised due to human activities over the centuries, particularly the uses that were held since the 1950s.

From a social point of view, the cultural-historical and economic factors that made the Mediterranean so prosperous also led to an unequal development of the different communities settled along its shores (King et al. 1997). The larger cities have become poles that have attracted more and more wealth and have increasingly enlarged their demographic basin: on the one hand, the smaller towns have suffered from progressive depopulation, while on the other hand, the large coastal cities have remained swamped by the growing volume of users that has fuelled a strong social demand for community spaces over time. It is therefore a priority to provide the citizens of such places with free and fair access to coastal resources, according to the principles of spatial justice. Similarly, climate change and increased urbanisation have generated a growing environmental risk from the sea, altering the safety of these spaces that are so crucial for urban well-being.

[1] This strategy is based on the concept that European maritime activities should be streamlined in a holistic manner, in order to achieve sustainable growth objectives (https://s3platform.jrc.ec.europa.eu/blue-growth).

1.4 Objectives and Structure of the Volume

The reasoning presented as an introduction to the present work has gradually defined the main objectives to follow, seeking to investigate whether current coastal planning and governance operate efficiently on coastal open spaces and public areas and whether the social and recreational functions of the shores of large contemporary cities are equally usable by citizens, considering their positive impact on human psycho-physical well-being and their relevant environmental value in deeply urbanised areas. It is therefore essential to address the issue of social needs in relation to the provision of proximity public spaces along urban coastlines. At the same time, the research is confronted with regulatory-legal aspects, as well as the theme of services provided by urban coastal areas, in relation to the health, environmental and socio-cultural offer; linked to this is the environmental risk, since phenomena such as coastal erosion, flooding, and sea-level rise influence the way of living the contemporary coast. These issues clearly fall within the macro-category of use and landscape management, key topic for urban studies and planning.

It is possible to break down this broad topic into more focused and targeted questions, to which a specific research approach is associated. The issue of whether coastal community users actually express a growing social demand for coastal recreational spaces will be addressed, using the in-depth study and comparison of coastal governance systems as a basis for investigation. It will be studied whether the ecological approach of contemporary planning can rehabilitate coastal social spaces, preventing climate change effects, through a literature review and the analysis of methods of ecological design applied to urban coasts. Finally, it will be analysed if a new integrated management-planning approach can respond to the changing socio-environmental demands of urban coastlines, through the study of local and international policies, directives and regulations, the collection of physical-spatial data and field interaction with stakeholders.

The articulation of the work envisages the division of this text into three main macro-sections. The first deals with the theme of the urban coast as a complex interface: in particular, it analyses the notion of *city-sea interface* as a design development potential for the sociality of urban coasts, and then describes the concept of the *urban coastal society* in relation to its connection with the *urban amphibious*, that is the flexible and adaptive declination of city coasts in a socio-environmental key, for the purposes of the integrated management of land spaces but also of maritime planning; finally, the socio-environmental crisis of contemporary coastlines in large cities is explored in relation to the ecological planning approach, and the theme of coastal governance within the Euro-Mediterranean basin. In the second macro-section, the programmatic and planning approach to coastal interface territories is introduced, proposing a methodological framework to support urban coastal planning and management: existing national and international directives and regulations on the topic of urban coasts and social uses of the sea in cities are analysed; after that, the definition of a methodological support to systematise the spatial, functional and geophysical data of urban coasts is described; moreover, the tools necessary to interpret the community perception of the city coastline, through social listening and field survey, are introduced. The third macro-section aims at comparing the urban coasts of Marseille and Naples, Mediterranean cities similar in size and population density, although different in their approach to coastal planning: the physical, legal, geographical and psycho-social characteristics of the two coastal interfaces will be investigated, applying the elaborated *methodological framework* to provide, in conclusion, an analytical basis to support planners and policy makers who regulate the transformation of urban shores.

References

Edward DB (2017) Marine ecosystem services. Current Biology 27(11): 507–510

Giannakidou C, Diakoulaki D, Memos CD (2019) Implementing a Flood Vulnerability Index in urban coastal areas with industrial activity. Natural Hazards 97(1)

Kay R, Alder J (2017) Coastal Planning and Management. CRC Press, London

Kim J (2015) Measuring the Equity of Recreation Opportunity: A Spatial Statistical Approach. Michigan State University, East Lansing

Kim J, Lyu SO, Song H (2019) Environmental Justice and Public Beach Access. City & Community 18 (1): 49–70

King R, Proufoot L, Smith B (eds) (1997) The Mediterranean: Environment and Society. Routledge, London

McElduff L, Ritchie H (2018) Fostering coastal community resilience. Mobilising people-place relationships. Area 50 (2): 186–194

Morena M (2011) Morphological, technological and functional characteristics of infrastructures as a vital sector for the competitiveness of a country system. An analysis of the evolution of Waterfronts. Maggioli Editore, Santarcangelo di Romagna

Pilkey O, Cooper JAG (2014) The Last Beach. Duke University Press, London

Pineschi G (2013) Pianificare la gestione delle acque e la tutela del territorio delle aree urbane: criticità e conflittualità. Qualità dell'ambiente urbano, IX Rapporto, Edizione 2013—Focus su Acque e ambiente urbano, ISPRA—Stato dell'Ambiente 46: 23–29

Porfyriou H, Sepe M (eds) (2017) Waterfronts Revisited: European ports in a historic and global perspective. Routledge, London

Small C, Nicholls RJ (2003) A global analysis of Human Settlement in Coastal Zones. Journal of Coastal Research 19(3)

Part I
The Urban Coast as a Complex Interface

Chapter 2
The City-Sea Interface and the Socio-Environmental Value of Urban Shores

Abstract The littoral urban zones and adjacent coastal marine space are interdependent subsystems, constituting a unified spatial, social and ecological system. Considering these dynamics is crucial for future transformations, aligning with rising expectations for quality of life and environmental accessibility. Investigating coastal settlement evolution and its impact on citizens is pivotal for informed socio-cultural, environmental, and urban planning decisions. For this reason, the chapter examines the city-sea interface, where the urban fringe meets the water fringe, forming a spatial and functional continuum. In particular, the attention is directed towards the ecological and socio-recreational potential of this peculiar coastal space.

Keywords City-sea interface · Proximity public spaces · Green–blue infrastructure · Ecosystem services · Ecological planning

2.1 The Concept of Interface as a Multifaceted Spatial Entity

Every place is characterised by specific physical, morphological and socio-cultural features that makes it in fact a unique and non-generalisable spatial reality, the result of the reciprocal relationship between material elements, personal and collective representations and the actions that its inhabitants perform to shape it: indeed, it is determined by a co-evolutionary process between the natural context and the communities that have experienced it over time (Becattini 2015). Regarding coastal areas, this is particularly true since this type of place is exceptionally peculiar, since it responds to heterogeneous identity traits in relationship with various kinds of spaces, uses and socio-environmental conditions. In order to study the processes at the base of the coastal socio-spatial and functional interactions, the concept of interface is introduced: it can be described as a linear or superficial geographical entity where two different systems come into contact; in general, original phenomena occur there, including exchanges of various kinds between the two parties, a reciprocal modification of the two original systems or even the exploitation of their differences and

I. Pistone, *Urban Coasts in Socio-ecological Transition*,
https://doi.org/10.1007/978-3-031-70783-4_2

peculiarities by business societies, cities or entire social communities (Brunet et al. 2005). It is thus a privileged zone of interaction between man and the environment, where human activities and functions are combined with the surrounding space. The interface is based on the relationships and juxtaposition of elements: interactions can take place within variable boundaries, as in the case of the contact between land and sea, and can sometimes be in strong competition between each other, as it happens for compact urban systems and adjacent natural environments (Lévy and Lussault 2019).

Since urban shores do not only refer to land areas, but also to water areas, it is possible to consider the coastal interface in relation to the reciprocal influence between the marine and land systems, exerted on both sides of the coast (Pittman and Armitage 2015). The concept of coastal interface is an important research topic for numerous disciplinary fields: it is in fact analysed in various scientific contexts that share the interest in defining the peculiarities of the interactions between land and sea and measuring their intensity, with the aim of supporting a responsible government of the territories that are most connected to the dynamics of urban water. This interest is clearly linked to the complexity of the relations between the different interpenetrating systems composing the littoral zones. Within the European context, it is estimated that the approximately 90,000 km of continental coastline are home to half of the entire EU population, while producing most of the European Union's wealth (European Commission 2001); it is clear that this generates a constant pressure that limits the quality of life, especially in the most urbanised areas, and that requires adequate integrated management for a sustainable use of the immense economic, environmental and social potential of the coastal space. From a geographical point of view, the interface is akin to a flexible area that facilitates spatial concentration and, consequently, greater human density. This space can be recognised from within and outside its variable boundaries, even unconsciously, through the social networks that are established there and on the basis of the social capital[1] that develops and distinguishes it. Thus, the social value shapes and defines physical space: its form and extent are characterised by a high degree of lability, changing over time according to the evolution of socio-functional relations. It can be compared to a geometric abstraction and studied according to the network of its inhabitants; in relation to the spatial distribution of its socio-environmental elements, the coastal interface may be analysed in a more physical sense, in order to understand social needs and the innovations they require (Nuvolati 2011).

Along the coastal strips, settlements present complex articulations: urban areas with variable density can be concentrated in accordance with a polar pattern to answer to the demand of resources, or they can be placed along the main road infrastructures parallel to the coastline. All this clearly entails more or less localised pressures on the socio-environmental ecosystem, generating spatial and urban morphologies that are often difficult to codify and manage (Abbate et al. 2009). The *land-sea*

[1] The concept of social capital generally refers to the wealth of relationships, experiences and values that an individual builds up in a lifetime in function of a specific place and society in which he or she fits (Andreotti 2009).

interface seems to be easy to define in the mind of researchers but it has an uncertain spatial delimitation, since it is the subject of broad urban studies, involving economic, geological, historical, cultural, ecological and engineering aspects. It is an area of correlation: its definition is in fact linked on one hand to the surface area between high and low tide, and on the other to complex economical, functional and legal evolution of the coastal and the water spaces (Zunica 1986). This interface takes into account the macro-territorial scale to analyse maritime space, increasingly fractioned by different stakes and the micro-local scale towards which land governance must tend in order to better manage the social demand.

From a physical and ecological point of view, the *coastal interface is* configured as a point of contact between land, sea and air, within a context of mutual environmental influence between the body of water and the land; it is a peculiar place since it hosts most of the biological cycle of the coastal ecosystem species (Della Croce et al. 1997). The resulting interactions that characterise this type of interface define specific areas, among which zones with residential, economic, productive, tourist-recreational and logistics vocations take place.

In particular, the *port-city interface is* characterised by its multidimensional landscape and its strong imprint on the surrounding territory (Fig. 2.1). On the other hand, if this kind of interface is the result of the mutual relationship between two entities (the city and the port), the intensity of this link depends on the degree of proximity of the two elements, which has evolved over the years according to urban

Fig. 2.1 The port-city interface shows the close relationship between the logistical infrastructure and the urbanised zones in which it is comprehended. In the photo, the area of the port of Catania (picture by the author)

and spatial changes. The port-city interface is also characterised by the presence of a distinctive architectural artefact that has undergone international influences, as a result of economic and commercial traffic, and could be integrated in the urban system as a public space, enjoyable by the community regardless of its primitive logistical function (Moretti 2020). In this sense, the urban-port context offers the possibility of experimenting with resilient planning in order to overcome past design criticalities offering new solutions that allow for more efficient and inclusive governance models to connect to the edges of the interface. This is important because the coastal interface in general has a limited land resource, consequently port design should provide for coexistence with urban-littoral values rather than their spatial relocation (Bruttomesso and Alemany 2011).

However, the commercial evolution of urban shores and the progressive decommission and transformation of former logistic areas since the end of the last century have generated a chaotic condition in which the aforementioned interface currently finds itself, progressively moving away from its role as the main gateway to the sea: the basic idea is that the port is a separate element from the interface, whereby the qualitative improvement of urban life is independent of the adequate planning of the city in relation to its harbour component (Ducruet and Notteboom 2022). In any case, it should be noted that there is a certain degree of port independence, as the harbour is in fact a system open mainly towards the sea from which sources of economic and cultural enhancement can be received.

The littoral interface is thus characterised by a multidimensional definition, difficult to categorise because it incorporates several definitions and numerous interpretations, depending on its specific uses and morphological conformation.

2.2 The City-Sea Interface: The Relationship Between Sea, City and Community

Among the different meanings of the littoral interface, this work aims to explore the concept of the *city-sea interface*, that is the area of physical, ecological, social and functional contact between the socio-ecologically structured city fringe and the water fringe, connected to the urban system close to the shore: this interface extends along highly urbanised land areas and along heavily used marine areas, constituting a relevant spatial and functional continuity (Fig. 2.2). At the basis of this concept, there is the hypothesis that the portion of the city closest to the coastal water is the urban area most influenced by the maritime element: the littoral urban fringe and the coastal marine space in close proximity to the shore are strongly interconnected subsystems that form a single spatial, social and ecological system. It is essential to consider them as a whole when considering their future transformation (Robert 2019).

In fact, its structure is not fixed but changes over time, in accordance with the social and environmental variations that the urban shore undergoes; what fuels the

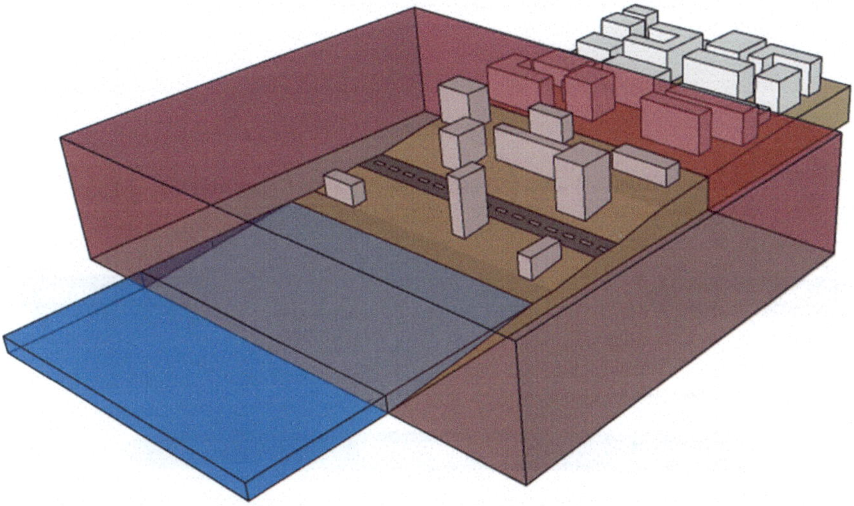

Fig. 2.2 Conceptual scheme of the city-sea interface, which hypothetical surface on land and in the sea is highlighted in red (elaboration of the author, based on Robert 2019)

transformations are the human, logistical, functional and environmental flows that characterise the margins of the two coastal systems and that allow for the effective articulation of the interface: the latter thus assumes a unique geographic and socio-spatial identity that is distinct from the basic terrestrial and aquatic domain. It is understood that the city-sea interface is characterised by relatively reduced spatial dimensions on both sides of the coast. Depending on the reference city and its urban configuration, the dimensions of the city-sea interface may of course vary, but, in any case, they will not include the entire width of the coastal area: for this reason, this concept can be mainly appreciated in large urban centres (Ollivro 2016). Within this particular urban area, three interrelated theoretical subsystems can be distinguished, namely the marine ecosystem, the urban ecosystem and the socio-system, based on the conceptualisation of the coastal system proposed by geographer Corlay (1995). The marine ecosystem is characterised by more pronounced anthropogenic activities than coastal areas further from the urban core; it is also more affected by the effects of urbanisation. The urban ecosystem, or geosystem,[2] has more sealed soils and greater demographic pressures. Finally, the socio-system is characterised by the prevalence of the social component that influences the entire coastal system also in an economic and perceptive sense (Fig. 2.3).

The socio-ecological relations between the three subsystems are more intense in the territory of the city-sea interface than anywhere else along the coast. Indeed, the interface is particularly sensitive to transformative phenomena such as sea-level rise

[2] The geosystem includes the relationships between physical and human elements, in antithesis to the ecosystem, emphasising the prevalence of anthropic processes over the natural (Lévy and Lussault 2019).

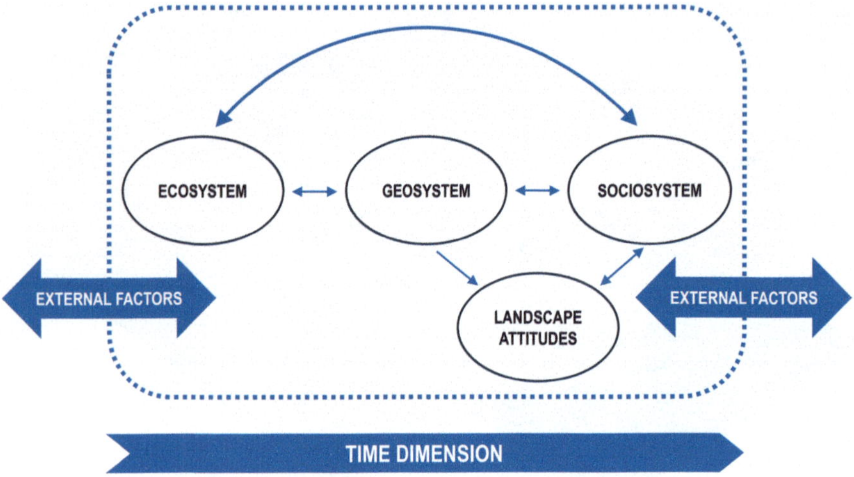

Fig. 2.3 The three subsystems underlying the concept of city-sea interface are mutually interrelated, from a spatial and time point of view (elaboration of the author based on Corlay 1995)

or globalisation: for this reason, climate change and ecological transition, intended as responses to large-scale changes, are particularly important for understanding and planning the city-sea interface. It is now well known that the reach and frequency of adverse weather events, from rising temperatures to disasters such as flooding, are increasing along the coasts of cities; this obviously has a tangible impact on the lifestyle and quality of life of the communities that inhabit the littoral areas, as their location is as privileged in relation to maritime benefits as it is critical with respect to risks from the sea (Siegel 2020). In any circumstances this criticality should be underestimated in urban planning and design, especially while setting up public uses and spaces, considering, moreover, the strong attractiveness of the coasts for numerous categories of stakeholders.

The importance of the social dimension of sustainable coastal development must be emphasised, especially for efficient planning. Although in large cities many uses of the shore can inspire transformative actions, the recreational and social perspective can be effectively integrated with ecological considerations, focusing on specific points of the city-sea interface, such as urban beaches or piers, which are important components of the urban coast as spaces intended for sociality. As a matter of fact, from a socio-recreational point of view, the city-sea interface represents an area of high frequentation and considerable interest, especially in recent years when the flow of coastal users has increased considerably (Evrard 2014). Given its relevant naturalistic and environmental component within highly urbanised contexts, it can attract citizens even far from the urban sea: the city-sea interface thus becomes a social platform for recreational uses of the coast, generating undeniable positive effects on human health and well-being. Since it is defined by the perimeter of the waters washing the city, the coastal interface is strongly linked to the theme of recreational access to the blue space: even if the area of contact with the sea are narrow, users are equally willing to travel considerable distances to enjoy the maritime resource, despite the fact that the availability of littoral public space is not always sufficient (Elliott et al. 2020).

It is a place where it is possible to make use of important landscape, cultural and environmental services, through social practices that can serve as an indicator of interactions between society and the coastal territory, considering specific land uses and the perception of place. However, the city-sea interface raises questions about socio-environmental equity and spatial justice, perfectly framed in the perspective of sustainable development: although the social and recreational function is only one of the many dimensions of integrated coastal zone management, at the present time social uses can be considered as a key to reconsider the way to analyse and design the shore in large cities, ensuring social balance, traffic reduction, innovative solutions and, in general, a stronger urban identity. Urban coastal zones are in fact places of great expectations: they are perfect spaces to conceive actions aimed at raising the quality of urban life and coherently organising traditional and recent uses. The development of coastal settlements often follows a morphological model intended to maximise the landscape enjoyment of unique places and produce physical transformations along the coastline in order to exploit its direct and indirect accessibility: thanks to its position, the city-sea interface fuels in fact the desire to be as close as possible to the sea. Obviously, in terms of economic growth, in this particular urban zone, the real estate value tends to increase proportionally to the proximity to the sea, often sacrificing free areas destined for the community in favour of building development: this pattern of growth tends to have a significant impact on the more natural parts of the urban shore, as well as on the cultural heritage and social fabric of the communities that inhabit it (Green 2010).

In the light of these reflections, it can be justified the choice to further explore the notion of city-sea interface as a resource for citizens, from both a social and necessarily ecological and environmental perspective, focusing the field of research on the spaces intended for public use along the coasts of large cities. In order to study the social function of the coast, one of the main objectives is to understand the practices and desires of coastal communities: the hypothesis is that nowadays the expectations of inhabitants tend to increase in terms of quality of life and accessibility to environmental services. It is therefore crucial to investigate the dynamics linking the coastal evolution of settlements and their potential for citizens, which should be the basis for interventions and decisions concerning socio-cultural aspects, environmental factors and anthropic elements of the city-sea interface (Fig. 2.4).

Fig. 2.4 Fruition of the city-sea interface in Naples: environmental amenities, recreation, sociality (picture by the author)

2.3 The Width of the City-Sea Interface: Land, Water, Islands

Flexibility is one of the main characteristics of the city-sea interface, as its conformation is directly dependent on the morphology of the coastline and the specific uses that users carry out there. On the terrestrial margin, the space is urbanised and follows the organisational structure of the city model, influenced by peculiar social and ecological features. On the marine margin, the ecosystem is in turn profoundly conditioned by the presence of the city, which influences its physiology, its functionality and obviously its evolution. The book attempts to conceptualise these mutual influences from a dimensional point of view: it is therefore proposed to schematise the city-sea interface in three bands, or edges, which are consequential to each other and possess a variable width: the *sea edge*, the *maritime city edge* and the *extended city edge* (Fig. 2.5).

- The *sea edge* indicates the area of the stretch of water closest to the shore. Its extension is linked to the geomorphological structure of the coast, to the presence of reefs, depressions, sandy seabed; the inner limit of this strip corresponds to the place where the waves break, moreover, generating currents and turbulence: for this reason, its surface is extremely variable in time and space and is bound to the depth of the marine floor (Short 1999). Indeed, for the purposes of a socio-recreational use of the interface, it is mandatory to consider the limits imposed by bathing safety or the technical characteristics of floating installations (such as platforms or similar elements).[3] In this sense, the uses of the sea are linked to the possible risks that evolve over time in relation to the environmental component; consequently, the definition of the sea edge must take into account the availability of data about the spatial planning of the body of water, the innovative methods of

[3] The European Union has drawn up UNI 11745:2019, which defines the area designated for bathing within a radius of 1 km, confirming the area of so-called *safe waters* with a depth of 1.60 m and signalled by special boats.

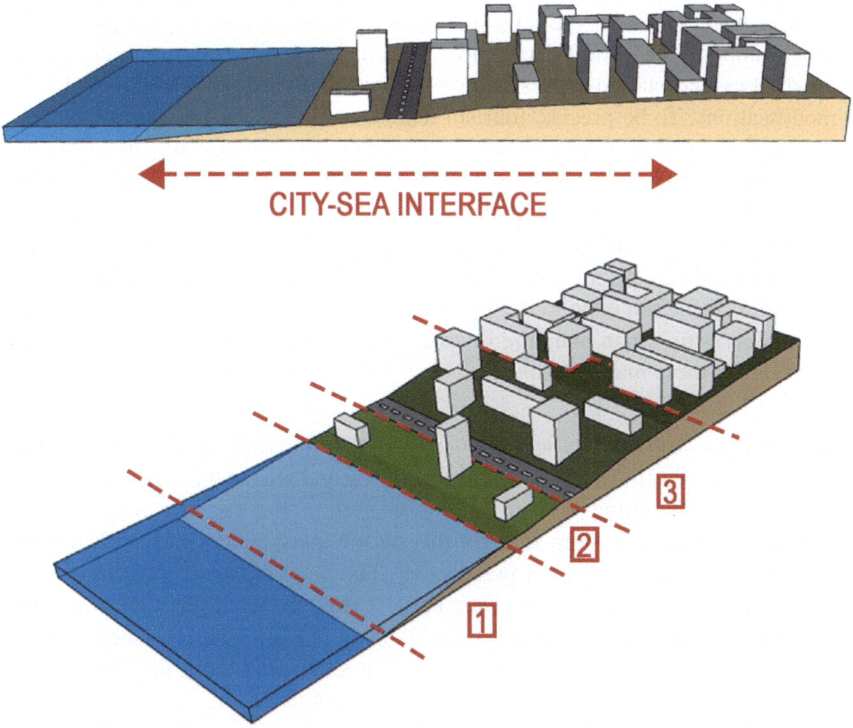

Fig. 2.5 The various edges of the city-sea interface: 1. Sea edge, 2. Maritime city edge, 3. Extended city edge (elaboration of the author)

meteorological marine forecasting and the probability that adverse climate events may modify the ways to exploit urban waters: a meta-dimensional perimeter of the analysed edge thus emerges (Simonetti 2019). Clearly, other activities are also concentrated in this sea space: suffice it to think about the water corridors for boats and ships to enter into the sea; some maritime areas are intended for various kinds of fishing, consequently dealing with rules for environmental and wildlife protection; seabed can be the subject for guided tours and it can welcome underwater art installations. Therefore, the sea edge is strongly influenced by local regulations and directives that expand or reduce its width depending on various factors, such as safety criteria, environmental characteristics or administrative and management delimitations. It is also important to consider how much the urban littoral conditions the island systems within the range of influence of the mainland: indeed, islands constitute a fundamental element in the careful management of the interface. In particular, the connection between the sea edge and the islands is inversely proportional to the distance of the latter from the urban coastline: the shorter the distance, the greater the dependence the island will have on the reference city on the mainland, as the urban system of the island will plausibly be configured as a satellite element of the urban pole beyond the body of water (Calado et al. 2007). Since the twentieth century, the progressive urbanisation of the littorals has rapidly altered the spatial configuration of islands nearby large coastal cities, resulting in loss of biodiversity and socio-ecological modifications. To be precise, tourism represents one of the most serious threats to island sustainability, along with environmental criticalities: it is plausible that a smaller urban system is less efficient in tolerating these issues (Leatherman and Beller-Simms 1997). However, the tourist industry itself can be considered as a source of wealth and an impulse for socio-cultural growth of smaller islands, as it constitutes a strong connection with the city-sea interface, expanding the perimeter of the sea edge to the point that these places become an integral part of it, addressing urban planning towards choices that emphasise the anti-fragile character of these maritime spaces (Nurse et al. 2014) (Fig. 2.6).

- The *maritime city edge* lies between the sea and the first road infrastructure parallel to the coast, called the *transit road*. Being the central area of the city-sea interface, it acts as an inner hinge in the urban-coastal system. It is the strip of coastal land closest to the water system and, for this reason, particularly important from an environmental and ecological perspective. The main *Nature's Contributions to People* are concentrated there (Díaz et al. 2018), for example, the effects that the local ecosystem, biodiversity and ecological processes have on the quality of life of users, in the form of coastal ecosystem services but also negative outputs (consider the effects of climate change that are amplified in this zone so close to the

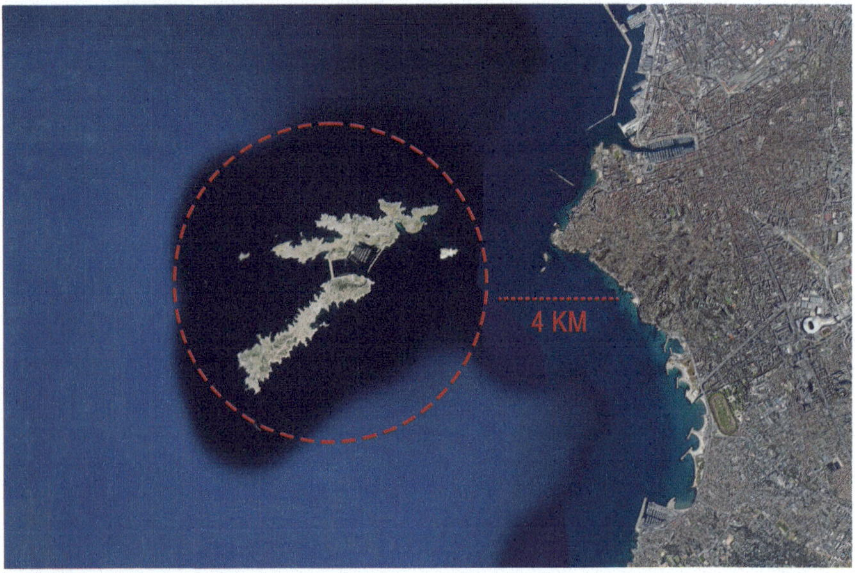

Fig. 2.6 The Frioul archipelago is extremely close to the Marseille coast (about 4 km away): it can be actually considered a peripheral part of Marseille within the sea edge (elaboration of the author)

water). If the coastal interface is in fact the area of reciprocal influence between the land and the sea, the maritime city edge stands as the stretch of coastline where the main biological cycles of species inhabiting this spatial environment take place (Tanacredi 2018). In this area, it is assumed that there is a more linear ecological continuity, due to the close proximity to the water element, which also has important implications on the development of social activities: since water itself is a physical factor that defines peculiar organisational and landscape patterns within coastal settlements, it is connected to specific littoral morphotypes at the base of the analysed portion of the city-sea interface. In particular, we consider trajectories orthogonal to the coast, determined by rainwater outflow paths that generate convexities and concavities: along these paths, it is possible to identify roads passing through the maritime city edge to connect the inner land to the coastline without abruptly interrupting its environmental continuity. At the same time, we consider morphogenetic trajectories parallel to the border between sea edge and maritime city edge: this category includes firstly the coastline itself, with its more or less sinuous course, depending on the conditions of the context, and secondly the roads parallel to the coast (such as the transit road). The last ones act as a gateway to the coastal environment but also constitute the first effective strong limitation to the littoral ecosystem, interrupting its continuity (Mininni 2006). Since the extent of the maritime city edge is determined by the actual distance between the coastline and the transit road, it is possible to state that its width is extremely variable: for this reason, there are city-sea interfaces endowed with a very wide maritime city edge (for example, settlements that rise really close to the water), as well as other interfaces where the maritime city edge is reduced until it completely disappears, as in cases where transport infrastructure run along the coastline. Although the presence of built areas in this zone may be of crucial importance in the research perspective, the maritime city edge is therefore particularly interesting for the potential presence of public or semi-public open spaces in a very close spatial and visual relationship with the body of water (Zhu 2020). Whether we refer to artificial shorelines, shaped by docks and piers that are not dedicated to logistic or productive functions, or to natural shorelines (for instance, those stretches of accessible coastline that are mainly characterised by beaches or reefs), the maritime city edge appears to be strongly defined by a morphological and socio-environmental point of view, hosting, at least in principle, a considerable amount of public areas in coastal urban centres that have become increasingly denser towards the sea (Fig. 2.7).

- Finally, the *extended city edge* is the most extreme part of the city-sea interface. It lies beyond the transit road and may extend more or less widely towards the innermost part of the urban coast. It is a zone of the interface with a fluctuant geometry and variable size, since its interest for the urban planning and design of socio-recreational public spaces on the urban coast depends on the presence of relevant green areas or open spaces connected to the coast, as well as on the housing density of the area, which should be extremely low. Its conformation is linked to the phenomenon of urbanisation that has gradually led communities to concentrate along the coastal strip and in the land in its immediate proximity,

Fig. 2.7 Marseille users enjoy physical contact with the sea from the pedestrian space of the Vieux Port, within the maritime city edge (picture by the author)

starting in the Seventies; in the last decade of the twentieth century, the shore already appeared almost completely built-up. It is only in the beginning of the new millennium that urban growth has progressively slowed down, although littoral zones are now compromised or at least rigidly characterised by major infrastructures influencing local ecosystems and a dense urban fabric invading much of the coastal territory (Salvati 2012). These phenomena are amplified by the need to accommodate new users, to satisfy socioeconomic needs, but above all by the impulse of tourism, a powerful factor in the artificialisation of the coastal landscape. However, the importance of open spaces in the extended city edge is measured in socio-environmental terms. Firstly, these areas play a crucial role in protecting the spaces closest to the shore from environmental risk phenomena that could cause huge human and economic losses: for example, the presence of vegetated areas in perfect condition can protect transport infrastructure and coastal spaces from landslides, floods and other hydrogeological impacts; they can also shield open spaces from extreme heat waves or torrential rains, increasing the liveability of the coasts to the communities that are settled there (Munang et al. 2013). Secondly, the open spaces of this zone can provide a design continuation for the public areas of the contiguous maritime city edge: in this sense, it is possible to give value to the large sites that are still vacant, albeit their number is quite limited in large contemporary cities, and to the small fragments of natural space between the buildings or in degraded areas. According to this vision, the extended city edge can be a suitable place for the enhancement of all the complex human-nature interactions that characterise the maritime city edge. The actual implementation could be operated through interventions that strengthen the sociality and ecological character of the open spaces in the innermost parts of the city-sea interface: the planning and management of these areas can indeed modernise their receptive, social and environmental capacity, offering communities new coastal services and generating new inclusive landscapes (Fig. 2.8).

Therefore, the description of the three edges composing the city-sea interface suggests that its structure can vary greatly, not only from city to city, but also within the same coastal urban area: its wider or narrower dimension depends on the littoral morphology that defines its spatial character, but also on the presence of very important social uses and open spaces of various kinds, to which users aspire for their uniqueness within the urban system.

Fig. 2.8 The park area of the Villa Comunale in Naples can be included in the extended city edge, as it represents an open public space beyond the transit road, within the frame of the city-sea interface (picture by the author)

2.4 Proximity Coastal Public Spaces for Green–Blue Infrastructure Project

The study of urban evolution allows to recognise the dependent and independent variables of the littoral landscape. In dimensional terms, coastal space is fundamentally indeterminate, an area where productive, economic and social functions and systems are free from material constraints and urban fragments gain value in the perspective of a continuous process of morphological modelling, expansion and retraction.

The city-sea interface represents a resource from many points of view, although it presents an articulated and not always homogeneous structure. In fact, its complexity also depends on the variegated elements that compose it: they can be categorised into a range of juxtaposed spaces, characterised by their greater or lesser aptitude to urban transformation.

Littoral areas characterised by a denser and more stratified urbanisation are certainly less flexible from a planning point of view, since the functionality of spaces is now consolidated and existing building has often saturated the available land: the result is a peculiar spatial configuration with respect to the adjacent maritime system. In this category, it is possible to include dense urban areas (for instance, residential areas or ancient cores of the settlements along the urban coastline, with deep historical, cultural and landscape values) but also industrial and port areas, in which the productive and logistical function often limits the free use of the coastal interface, as the economic weight of these places is greater than the social desire that communities can manifest towards the urban sea.

On the other hand, the areas of the coastal interface most prone to transformation are those that show either a higher degree of naturalness or a weakly defined, or even fading, functionality. These include unequipped or lightly equipped natural spaces (such as urban beaches), urban parks of various size along the coastline (in this case, it is important to consider the landscape constraints that may influence the design process for these areas), socio-recreational areas at a local and micro-local scale (for example, small residual spaces in the urban fabric that are often used as meeting places for market functions, or for the spontaneous installation of street furniture), up to abandoned, degraded or disused areas, as well as public places perceived as unsafe by users: even if these place do not present a high ecological value, they may have a discrete functional flexibility from a planning and operational perspective. The elements of the latter category, in particular, can be read as an opportunity for sustainable social development, as potential components of a multifunctional green–blue infrastructure, in relation to ecological and landscape approaches (Di Venosa and Manigrasso 2022) (Fig. 2.9).

The third objective of the *Sustainable Development Goals* states that it is necessary to promote community well-being to ensure sustainable development: coastal planning should focus on the physical and mental health of users, on their satisfaction and on guaranteeing a feeling of positive control over their surroundings. Moreover, several studies show that the triggers for cancer, heart and cerebrovascular problems and respiratory disorders can get worse because of an overly oppressive and

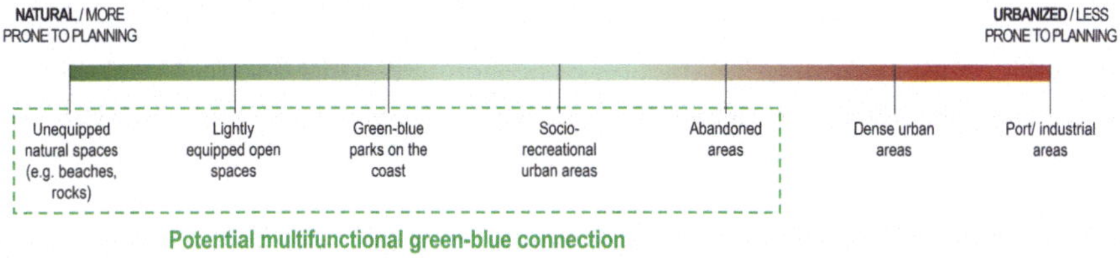

Fig. 2.9 The spaces of the city-sea interface may have greater or lesser flexibility and predisposition to design change, depending on their degree of naturalness, among other things: in this sense, it is possible to hypothesize the functional reconnection of public spaces through ecological planning approaches, such as green–blue infrastructure (elaboration of the author)

unhealthy built environment and sedentary lifestyles. In addition to this, public health is increasingly threatened by external causes such as climate change, which is seen as one of the most compelling challenges of urban communities in the new millennium (Andreucci et al. 2019).

Green–blue infrastructure is an urban planning approach related to nature-based strategies and design solutions and to the reticular development of space: in fact, it can be defined as a strategically planned network of natural and semi-natural areas, whose environmental characteristics can provide a wide range of ecosystem services (Ghofrani et al. 2017). It is also related to the concept of natural capital that every urban area possesses: its founding elements are integrated with the built environment, through different design systems, such as permeable pavements and rain gardens, fostering environmental protection, citizens' wellbeing, social equity and inclusion: it is therefore a hybrid infrastructure composed by green and blue public spaces (in case water-coastal ecosystems are included) and more urbanised areas, comprehending both extensive zones, like parks, urban forests, wetlands and coastal creeks, but also punctual and spatially delimited elements, such as public gardens, narrow community open spaces, depaved areas, ponds and small lakes, urban beaches (Andreucci 2017). Due to its inherent complexity, littoral green–blue infrastructure involves multi-scalar and multi-actor decision-making processes in order to preserve urban-coastal ecosystems from degradation by increasing their adaptive capacity to global changes. This kind of approach can be included in the field of ecological planning: its rehabilitative action focuses on those areas that are particularly critical and neuralgic in the urban-environmental system in order to regenerate vital ecosystem connections and services for local communities, translating planning interventions into more or less strict policy measures, on a case-by-case basis (Chavez et al. 2021). *Nature-based Solutions* (NbS) absorb this philosophy, in order to manage contemporary challenges in a sustainable manner, even at the small scale along the city-sea interface, and restoring ecosystem nature with an adaptive approach. Simultaneously, they provide benefits for human well-being and biodiversity, in a flexible combination with other environmental, technological or engineering solutions aimed at solving socio-environmental issues: great attention is given to the specificity of intervention sites, even at the local scale, considering their community and cultural characteristics. They are tools designed to grant users an equal accessibility to coastal resources, promoting transparent and participatory planning and recognising trade-offs between the immediate production of economic benefits and the development of various ecosystem services over the long term (Cohen-Shacham et al. 2016).

It is evident how important it is to take these specific characteristics into account in the reorganisation of the public spaces of the city-sea interface, as a platform for the socio-recreational connection of its community, rethinking their functional meaning and facilitating their management and fruition, in accordance with the current social and recreational need that is less oriented towards large public spaces dedicated to tourists and occasional visitors: instead, the actual social demand asks to emphasise high quality proximity public spaces in order to allow citizens to experience and come into contact with the sea and the coasts of their city (Pittaluga 2018). It therefore seems fundamental to study the coastal interface in function of the opportunities it offers to its main users who experience it on a daily basis and express a desire for sociality: in this regard, the concept of the *15-minute city* fits in. This theory proposes an urban model based on a small-scale and extremely localised planning, rooted in the hypothesis that the primary functions that society needs, from a productive but above all social point of view, can be located within a limited radius of action, rounded down to approximately 1.5 km, representing the average distance that can be covered by foot in precisely fifteen minutes. The approach addresses issues related to recreation and sociality, proposing the realisation of high-quality neighbourhood public spaces that are equally accessible to the different segments of littoral society, but also prone to sustainable urban development. The reduction of the influence range of functions related to sociality could ensure higher quality standards at the neighbourhood level and become part of the ecological transition process (Moreno 2020), limiting travel outside dense urban centres to enjoy the benefits of the sea.

Proximity thus becomes a systemic character of the network of coastal open spaces, composed by places but also interactions between individuals, functions and the environment, according to which users tend to form social relationships on the basis of a closeness criteria (Manzini 2021): this concept can be a spur for mending the fragmentation inside communities living along the coast, but also for recalibrating the relationship with the physical-environmental aspects of the city-sea interface. In fact, the coastal city, and consequently the urban shore, can be described as a macro-ecosystem on a territorial scale within which local micro-ecosystems are articulated: the latter are connected to each other as minimal ecological units that can achieve a certain functional independence while maintaining links with adjacent units. Recomposing these micro-ecosystems can push city-sea interface planning towards a more rational management of resources and public spaces, also with regard to protection from environmental criticalities.

In summary, contemporary city-sea interfaces must prove to be safe, efficient, but also accessible, providing better services and increasing the quality of life of local society, through a careful design of the public spaces along the coastline, especially in the areas of the city that are deeply connected to the urban sea: today, these spaces are often largely congested and scarcely enjoyable, as is well known. In this sense, resilience is a primary character for coastal areas, in order to respond adaptively to the adverse circumstances that could conceivably affect their most sensitive spatial spheres: the ability to face environmental, sociological and functional changes for the purposes of regaining a balanced status is necessary for the littoral interface when it comes to deliver efficient services and development opportunities to the coastal community. Therefore, one of the main objectives of coastal ecological planning should be healing the damage suffered by the urban-littoral fabric, so that a reorganisation of the fragmented structure of the city-sea interface can take place, restoring a sustainable spatial wholeness for its inevitably anthropized context: the systemic nature of the green–blue infrastructure aims to coherently respond to multiple demands expressed both by the city with its inhabitants and by the coastal environment itself. As mentioned above, this approach aims, in fact, at increasing the quality of life of citizens through social and economic actions based on the correct multifunctional use of environmental resources. This makes it a potentially valid policy tool since multifunctionality could positively contribute to the satisfaction of the needs of various stakeholder segments, promoting social growth, dealing with the equal use of coastal public spaces and supporting the development of an efficient green economy for an effective management of the coastal territory (Veerkamp et al. 2021).

2.5 The Territory of the City-Sea Interface for a Potential Socio-Environmental Innovation

Within the context of the city-sea interface, it can be stated that land and sea are intrinsically connected through multiple and complex socio-ecological interactions of paramount importance in the economic subsistence of coastal communities, but also in terms of the social well-being to which these communities aspire. The relationships between the terrestrial and water systems could be divided into three categories (Álvarez-Romero et al. 2011). In the first one, there are the flows that underlie the unfolding of socio-ecological processes, for example, the watercourses running into the sea, the migration of certain species depending on the time of year, but also the progressive groupings of communities around polar elements, such as coastal areas that undeniably provide resources and strong social cues. On the other hand, the second category groups elements that disturb and threaten the systemic equilibrium, such as the alteration of habitats or the change in the morphology of places: these critical issues are differentiated into exogenous risks, like climate change and coastal erosion linked to sea-level rise, and into endogenous risks, like the disconnection of ecological networks due to the urban fabric. Finally, the third category includes aspects that are related to management and operational actions regarding the positive and sustainable development of interactions within the coastal interface, in relation to the areas to be protected, the functions to be assigned to coastal spaces or the adoption of restrictions on land (and water) use: such choices must pursue beneficial effects on human lives and return positive outcomes to the coastal environment, thus falling within the scope of territorial-environmental governance (Antonioli 2017), that is the decision-making processes that society and its representative institutions undertake, affecting the status, functionality and accessibility of the city-sea interface.

In this sense, the phenomenon of coastal urbanisation can be observed as much on land as in the sea: the effects can be measured in terms of ecological and environmental impacts, due to various forms of pollution, but also to urban expansion on the water, in relation to interventions to increase the walkable surface area of the city (Crossland et al. 2006). It is possible to find a condition of heterogeneity in urban shorelines that are characterised as liminal landscapes between sea and city, the site of physical, biological and anthropic interactions that produce original environmental contexts in which different kinds of elements share mutual relationships (Phillips 2005). It can be said that city shorelines simultaneously assume the role of interruption and connection between the land and water margins, representing both a potential for development and a physical-social obstacle to the evolution of the city (Pavia and Di Venosa 2012).

From a functional point of view, the city-sea interface is thus configured as the habitat peculiar of coastal community in which each user carries economic, but also residential, social and recreational demands, shaping the coastal ecosystem, or in any case setting forth precise expectations regarding its transformation, and attributing a particular perceptive character to the shore (Turri 2002). It is a part of the city that could be considered as a territorial entity rather than a mere place, as a consequence of the specific social practices generated by the reticular relations involving both human societies and environmental and functional elements in the transformation of physical space (Magnaghi 2020; Painter 2010). This reconnects the coastal interface to the bio-cultural character of the landscape: the area of the territory as perceived by local communities is the outcome of interactions between human and natural factors, including both valuable spaces and ordinary or degraded places, in the natural and urban spheres. Hence, great importance is given to the identity and perceptive value. It allows people to understand and interpret the structure of the territory, in the light of its strengths, its weaknesses and, above all, its antithetic features breaking down its conceptual order and static organisational homogeneity (Minca and Colombino 2012). The city (especially in its most critical and complex areas, such as the urban shore) tends to fragment and reassemble itself in a continuous and dynamic spatial reorganisation that must be thoroughly understood. The city-sea interface is in fact characterised by a mutual interdependence with functional programmes: because of this, it is appropriate to talk about a fluid spatiality that that constantly reshape its own structure, although this articulation is not always grasped in the relationships between sea and coast, especially in policy-making and planning (Morri 2012; Soriani 2002).

The city, regardless of its size and number of inhabitants, is an aggregation in which diversified and complex dynamics take place. On the other hand, territory is considered an asset with its own identity that binds itself to the city reality with linear and punctual elements, through tangible and intangible signs, going beyond the generic definition of enclosure for the urban organism. The relationship of continuity between the urban and territorial entities converges in the notion of landscape. For this reason, the urban system crystallises characteristics of the territory linked to history, memory or even future development, but it cannot be considered as a random spatial representation of human settlements. Instead, it can be described as a multifaceted state of the territory that manifests with greater density the continuity of the territorial structure, transcending the dimensional aspect. This concept can be extended to the littoral context of this volume. In short, the notion of territory is configured as a process of socio-environmental definition of places, in this case, along the city-sea interface, in which the relations between coastal and urban systems transcend dimensional boundaries and change over time: in the same way, also the local society changes while enjoying coastal territory through affective-identity representations, specific uses and practices and peculiar resources (Banini and Ilovan 2021). Indeed, the city-sea interface provides a context within which the direct or indirect actions of users and the knowledge produced by experts in coastal studies must be delimited geographically, functionally and in terms of governance, effectively defining the urban-coastal space (Brun-Picard 2020). The territorial identity thus assumes an important value in the study of the city-sea interface because it condenses private and collective interests underlying the principle of belonging to place and the specific needs and expectations of users, towards which thoughtful planning should tend in order to guarantee adequate community use of urban coastal areas (Panzera 2022).

2.6 Sustainable Development Along the City-Sea Interface Between Urban and Maritime Planning

While studying the city-sea interface concept, it must be emphasised that land and sea uses, environmental and ecological characteristics and social demands are closely interconnected: despite this, interactions between land and sea areas are often ignored in local urban planning. The development of coastal areas is directly related to the sustainable use of the maritime space. A large number of activities take place at sea and at the same time are related to the coastal zone. These functions can be categorised into three main macro-categories, which are listed below: *maritime connections* indicate the movement of goods and passengers, but also operations for military purposes; the *exploitation of the sea resources* includes mining, renewable energy production, fishing and aquaculture; finally, among the *uses related to the natural environment fruition*, it is possible to group research and scientific dissemination and recreational and cultural uses (Smith et al. 2011). Since the coast is made up of land and water spaces on which functions often take place simultaneously, it is clear that conflicts can arise in the uses of the city-sea interface: this is particularly true for a context such as the Euro-Mediterranean basin in which economic-tourist sectors are increasingly developing, despite the existing functional frictions.

The coastal interface can be defined as an open system, in relation to the continuous interactions between the littoral land and the adjacent body of water: however, it must be considered that marine ecosystem tends to show greater openness than its terrestrial counterparts, as it is more malleable from a functional point of view. Indeed, there are no physical boundaries in the sea, just as functional zones are less clearly defined, at least in appearance, and give rise to easier, albeit damaging, overlap of uses. There is a need for efficient and coherent planning integration between the two systems within the urban coastal structure, overcoming any political and management barriers in favour of environmental well-being and equal usability. Moreover, relationships between the land and the maritime spheres occur on a wide range of spatial and temporal scales that vary according to coastal morphology and social context: the anthropic contribution usually alters these interactions and its effects must necessarily be interpreted in the long term (Stoms et al. 2005).

In the contemporary urban planning debate, it emerges that maritime spatial planning, which is historically more recent, tends to be conducted separately from land use, which is instead more rooted within traditional planning and management processes (Walsh and Kannen 2019). Responsibility for terrestrial and marine planning often falls on administrative bodies at different levels of governance, complicating coordination and integration between the two areas that are nevertheless two sides of the same coin (Morf et al. 2022). From this perspective, the notion of city-sea interface can be proposed as a key for understanding coastal and sea planning, as a connecting element between two inseparable and interdependent ecosystems: their delicate socio-spatial balance appears to be threatened by significant pressures from competing anthropogenic activities that lead to the environmental and functional degradation of coastal spaces (Tsilimigkas and Rempis 2017). The concept of the city-sea interface brings the urban planning field back to a reasonably local scale, interweaving spatial plans with community needs and ecological features, defining an operational link between two types of space (the land and the sea) separated only by the ephemeral boundary of the coastline.

References

Abbate G, Giampino A, Orlando M, Todaro V (eds) (2009), Territori costieri. FrancoAngeli, Milan

Álvarez-Romero JG, Pressey RL, Ban NC, Vance-Borland K, Willer C, Klein CJ, Gaines SD (2011) Integrated land-Sea conservation planning: the missing links. Annual Review of Ecology, Evolution, and Systematics 42: 381–409

Andreotti A (2009) Che cos'è il capitale sociale. Carocci Editore, Rome

Andreucci MB (2017) Progettare Green Infrastructure. Ipsoa, Milan

Andreucci MB, Russo A, Olszewska-Guizzo A (2019) Designing Urban Green Blue Infrastructure for Mental Health and Elderly Wellbeing. Sustainability 11(22)

Antonioli M (2017) Sostenibilità dello sviluppo e governance ambientale. Giappichelli Editore, Turin

Banini T, Ilovan O (eds) (2021) Representing Place and Territorial Identities in Europe. Springer, Cham

Becattini G (2015) La coscienza dei luoghi. Donzelli Editore, Rome

Brunet R, Ferras R, Théry H (2005) Les mots de la géographie: Dictionnaire critique. La documentation française, Paris-Montpellier

Brun-Picard Y (2020) La notion d'interface en géographie. Conceptualiser l'indissociable lien au support terrestre. L'Harmattan, Paris
Calado H, Quintela A, Porteiro J (2007) Integrated Coastal Zone Management Strategies on Small Island. Journal of Coastal Research SI(50): 125–129
Chavez V, Lithgow D, Losada M, Silva-Casarin R (2021) Coastal green infrastructure to mitigate coastal squeeze. Journal of Infrastructure Preservation and Resilience 2 (7): 1–12
Cohen-Shacham E, Walters G, Janzen C, Maginnis S (eds) (2016) Nature-based Solutions to address global societal challenges. IUCN, Gland-Cambridge
European Commission (2001), L'UE e le zone costiere. Sulle coste d'Europa spira un vento nuovo
Corlay JP (1995) Géographie sociale, géographie du littoral. Norois 42(165): 247–265
Crossland JC, Baird D, Ducrotoy J, Lindeboom HJ (2006) The Coastal Zone - A Domain of Global Interactions. In Crossland CJ, Kremer H, Lindeboom HJ, Crossland JC, Le Tissier MDA (eds), Coastal Fluxes in the Anthropocene. The Land-Ocean Interactions in the Coastal Zone Project of the International Geosphere-Biosphere Programme. Springer, Berlin, p. 1–37
Della Croce N, Cattaneo Vietti R, Danovaro R (1997) Ecologia e protezione dell'ambiente marino costiero. UTET, Naples
Di Venosa M, Manigrasso M (2022) Coste in movimento: Infrastrutture ambientali per la rigenerazione dei territori. Donzelli Editore, Rome
Díaz S, Pascual U, Stenseke M, Martín-López B, Watson RT, Molnár Z, Hill R, Chan KAA, Baste IA, Brauman KA et al (2018) Assessing nature's contributions to people. Recognizing culture, and diverse sources of knowledge, can improve assessments. Science 259 (6373): 270–272
Ducruet C, Notteboom TE (2022) Revisiting port system delineation through an analysis of maritime interdependencies among seaports. GeoJournal 87(3): 1831–1859
Elliott L, White MP, Grellier J, Garrett JK, Cirach M, Wheeler BW, Bratman GN, van den Bosch MA, Ojalah A, Roiko A, Lima ML, O'Connor A, Gascon M, Nieuwenhuijsen M, Fleming LE (2020) Research Note. Residential distance and recreational visits to coastal and inland blue spaces in eighteen countries. Landscape and Urban Planning 198
Evrard B (2014), La côte, un terrain de jeu? De l'utilitaire au récréatif. Presses universitaires de Rennes, Rennes
Ghofrani Z, Sposito V, Faggian RA (2017) Comprehensive Review of Blue-Green Infrastructure Concepts. International Journal of Environment and Sustainable Development 6(1): 15–36
Green RJ (2010) Coastal Towns in Transition. Local Perceptions of Land-scape Change. Springer, Cham
Leatherman SP, Beller-Simms N (1997), Sea-level rise and small islands. An overview. Journal of Coastal Research SI(24): 1–16
Lévy J, Lussault M (eds) (2019) Dictionnaire de la géographie et de l'espace des sociétés. Belin Editeur, Paris
Magnaghi A (2020) Il principio territoriale. Bollati Berlinghieri, Turin
Manzini E (2021) Abitare la prossimità. Idee per la città dei 15 minuti. Egea, Milan
Minca R, Colombino AF (2012) Breve manuale di geografia umana. CEDAM, Padua
Mininni M (2006) L'approccio dell'ecologia urbana. In Vitale A (ed) Ritrovare il mare. Linee guida per gli interventi di riqualificazione della fascia costiera di Napoli. CLEAN, Naples, p. 36–42
Moreno C (2020) Droit de cité: De la "ville-monde" à la "ville du quart d'heure". Editions de l'Observatoire, Paris
Moretti B (2020) Paesaggi Logistici e Infrastrutture Portuali di Confine. La nascita della Città del Cluster. Urbanistica Informazioni 289(6): 15–19
Morf A, Moodie J, Cedergren E, Eliasen SO, Gee K, Kull M, Mahadeo S, Husa S, Vološina M (2022) Challenges and Enablers to Integrate Land-Sea-Interactions in Cross-Border Marine and Coastal Planning: Experiences from the Pan Baltic Scope Collaboration. Planning Practice & Research 37(3): 333–354
Munang R, Thiaw I, Alverson K, Mumba M, Liu J, Rivington M (2013) Climate change and ecosystem-based adaptation: A new pragmatic approach to buffering climate change impacts. Current Opinion. Environmental Sustainability 5(1): 67–71

Nurse LA, McLean RF, Agard J, Briguglio LP, Duvat-Magnan V, Pelesikoti N, Tompkins E, Webb A (2014) Small islands. In Vv. Aa., Climate Change 2014: Impacts, Adaptation, and Vulnerability. Part B: Regional Aspects. Contribution of Working Group II to the 5th Assessment Report of the IPCC. Cambridge University Press, Cambridge, p 1613–1654

Nuvolati G (ed) (2011) Lezioni di sociologia urbana. Il Mulino, Bologna

Ollivro J (2016) De la mer au meritoire. Faut-il aménager les océans?. Editions Apogée, Rennes

Painter J (2010) Rethinking Territory. Antipode 42(5): 1090-1118

Panzera E (2022) Cultural Heritage and Territorial Identity. Springer, Cham

Pavia R, Di Venosa M (2012) Waterfront. Dal conflitto all'integrazione. LISt Lab Laboratorio Internazionale Editoriale, Trento

Phillips A (2005) Landscape as a meeting ground: Category V. Protected Landscapes/Seascapes and World Heritage Cultural Landscapes. In Brown J, Mitchell N, Beresford M (eds), The Protected Landscape Approach. Linking Nature, Culture and Community. IUCN, Gland-Cambridge.

Pittman J, Armitage D (2015) Governance across the land-sea interface: A systematic review. Environmetal Science and Policy 64: 9–17

Pittaluga P (ed) (2018) Insediamenti turistici costieri e sostenibilità. Progetti di rigenerazione. FrancoAngeli, Milan

Morri R (ed) (2012) Insegnare il mare. Paesaggi costieri e vocazioni marittime. Carocci Editore, Rome

Robert S (2019) L'urbanisation du littoral: espaces, paysages et représentations. Des territoires à l'interface ville-mer. Université de Bretagne Occidentale (UBO), Brest

Salvati L (2012) Sprawl e compattezza nei territori in transizione: verso una nuova urbanizzazione mediterranea. Memorie Geografiche 9: 273–290

Short A (1999) Handbook of beach and shoreface morphodynamics. Wiley, Hoboken

Siegel FR (2020) Adaptations of Coastal Cities to Global Warming, Sea Level Rise, Climate Change and Endemic Hazards. Springer, Berlin

Simonetti F (2019) Sicurezza della balneazione: la norma UNI 11745:2019. Studi costieri 28 (99)

Smith HD, Maes F, Stojanovic TA, Ballinger RC (2011) The integration of land and marine spatial planning. J Coast Conserv 15(2): 291–303

Soriani S (2002) Porti, città e territorio costiero: le dinamiche della sostenibilità. Il Mulino, Bologna

Stoms DM, Davis FW, Andelman SJ, Carr MH, Gaines SD, Halpern BS, Hoenicke R, Leibowitz SG, Leydecker A, Madin EMP, Tallis H, Warneret RR (2005) Integrated coastal reserve planning: making the land–sea connection. Frontiers in Ecology and the Environment 3(8): 429–436

Tanacredi JT (2018) Preserving Biological Diversity: Coastal Ecosystem Restoration Not In Balance. In Tanacredi JT, The Redesigned Earth. Springer, Cham, p. 29–52

Bruttomesso R, Alemany J (eds) (2011) The Port City of the XXIst Century. New Challenges in the Relationship between Port and City. RETE Publisher, Venice

Tsilimigkas G, Rempis N (2017) Maritime spatial planning and spatial planning: Synergy issues and incompatibilities. Evidence from Crete island, Greece. Ocean & Coastal Management 139: 33–41

Turri E (2002), La conoscenza del territorio. Metodologia per un'analisi storico-geografica. Marsilio Editore, Venice

Veerkamp CJ, Schipper AM, Hedlund K, Lasarova T, Nordin A, Hanson HI (2021) A review of studies assessing ecosystem services provided by urban green and blue infrastructure. Ecosystem Services 52

Walsh C, Kannen K (2019) Planning at sea: Shifting planning practices at the German North Sea coast. Spatial Research and Planning 77(2): 147–164

Zhu X (2020) Optimization design of public space structure in coastal cities based on interactive design. Journal of Coastal Research SI(104): 751–755

Zunica M (1986) Per un approccio con l'interfaccia terra-mare. Quaderni del Dipartimento di Geografia - University of Padua, Padua

Chapter 3
Urban Coastal Society and Its Relationship with the Sea

Abstract The functional relationship between the urban littoral of large contemporary cities and their inhabitants defines the urban coastal society, which is intimately linked to the area directly connected to the sea and to the coastal water itself. As the urban littoral and the urban coastal society mutually readjust themselves to environmental and social coastal issues, the city-sea interface can be described as an urban amphibious that serves as a hub for socio-economic and environmental services. In this context, littoral awareness becomes central to more efficient planning that takes into account the expectations of the community.

Keywords Urban coastal society · Coastal commons · Urban amphibious · Marine citizenship · Socio-environmental demands

3.1 Definition and Main Characteristics of Urban Coastal Society

The development of urban settlements along the coast has been influenced, over time, by numerous factors. Although spatial and socio-functional processes are highly articulated, it is nevertheless possible to briefly divide the elements underlying this development into three macro-categories: the geophysical aspects include features such as advantaged location and easy access to maritime resources; the economic aspects concern the richness of organic and inorganic natural resources present along the coastlines, the possibility of establishing successful import and export trades and producing energy from the sea and water, but also, in more recent times, the growing tourism industry; finally, among the landscape-environmental aspects, it is possible to mention the atmospheric benefits for the health and well-being of users, the immense cultural and natural beauty that characterises the coasts, besides their significant social and recreational value (Fig. 3.1). Based on these elements, communities settled along the coastal interface tend to grow exponentially, especially where coastal cities are denser. The global population, estimated at 7.6 billion citizens in 2018, is expected to rise to a staggering 9.9 billion by 2050: in the cities, there

I. Pistone, *Urban Coasts in Socio-ecological Transition*,
https://doi.org/10.1007/978-3-031-70783-4_3

Fig. 3.1 Phlegrean landscape from the Bagnoli pier, in Naples: landscape-cultural assets in a complex functional system (picture of the author)

are currently 4.1 billion of the world population and this number will grow to 6.9 billion, mostly distributed along coastal areas[1].

In relation to the theme of shores in large contemporary cities and the relationship established with the citizens who inhabit them and experience their changes, it is possible to define the *urban coastal society* as a community intimately connected to two specific elements. The first is the coast itself, intended as the urban area in direct contact with the body of water. The second element is the coastal water, which is essential in the process of ecological transition and environmental regeneration interventions: in fact, water cannot be considered as an element separated from the land, but rather as a component fluidly linked to the urban system whose coasts are shaped by everyday uses of people who formally or informally enjoy them (Hannigan 2017). The changing nature of city-sea interface therefore reflects the diverse needs and experiences of the community, requiring a high degree of socio-spatial flexibility. Indeed, the elements underlying coastal identity change over time but also in space, influencing the great symbolic power that coastlines possess: this makes difficult to univocally classify the components of littoral communities (Gillis 2012). The concept of place identity turns out to be fundamental in the study of the city-sea interface territory, conceived as a dynamic, open, porous place available for functional implementation: on the basis of these notions, the territorial identity of the

[1] Data available from the *World Population Data Sheet* (2018) of the Population Reference Bureau in Washington, DC.

urban coastal society is comparable to a process of social construction, which can be analysed through quantitative methods and analytical-planning tools involving various stakeholders and which is essential for the community to settle in a given place for its distinctive features, develop shared values, solutions, actions and future trends (Banini 2019).

The maritime city, and therefore its city-sea interface, can be considered as an expression of the identity of the coastal society that in different ways shapes it, generating a fusion of cultures, habits, uses and functions of the coast and the sea: it is possible to study it as a community habitat closely connected to the sea, within which different users from different social classes live and interact, sharing a collective coastal memory that is specific to each littoral interface (Clemente 2013). Historically, the evolution of urban coastal societies began with the identification of the logistical value of the coastline; subsequently, economic growth led to a socio-environmental coastal deterioration. Communities then rediscovered an interest in environmental and ecological issues, also due to the loss of identity-functional connection between the city, its society and the coast (Timur 2014). However, it is not easy to identify precisely and exactly the components of this unique kind of community. The urban shore represents a *unicum* because it is the gateway to naturalness within the contemporary urban fabric, it offers health benefits and allows specific uses, it can be fertile ground for economic investment and encourages sociality: this is clearly important for a wide range of users that goes far beyond the spatial boundaries of the city-sea interface (Fig. 3.2).

Fig. 3.2 Juxtaposition of coastal social uses at Esplanade J4 in Marseille (picture of the author)

Immobility in coastal communities is thus rapidly becoming incompatible with its variable sociological structure: further and further, the urban system is coming into contact with coastal areas closer to the sea. For instance, considering places that are far away from main city centres, the users and occasional visitors' pressure increases just in certain periods during the year (Smith and Thomsen 2008), while core areas of the city-sea interface are more evenly frequented over the months. The combination of people living nearby the urban coastline and large numbers of national and international tourists can result in significant population growth, along with frequent changes in community composition and structure.

Among the users of the coastal society, it is therefore possible to identify permanent residents of the city-sea interface, users who visit the shore to work or for recreation at different moments of the week, stakeholders involved in the tourist, real estate or commercial business, people from other cities, regions or even nations, but also political representatives, socially engaged associations, as well as additional figures who can enrich the sociological structure of the littoral interface (Schwermer et al. 2019). In the perspective of coastal interface management, it is plausible that the urban coastal society will never reach an absolutely well-defined and fixed status: on the one hand, this might be a critical issue to actively involve a significant number of users in coastal governance, considering that a large part of the individuals impacting the coastal area might reside elsewhere or might only have a short-term interest in the zone (Thomsen et al. 2009); on the other hand, it should be noted that the urban coastal society would not exist if there were no real involvement of its users in the fruition of the sea and the coast, which have always exerted a great appeal on the urban and peri-urban population.

Especially in larger cities, urban coastlines gather socio-economic values and provide a varied range of urban services: housing functions, health benefits for citizens, socio-cultural activities, transportation and job opportunities combine with the numerous ecosystem services produced in these complex urban environments, such as good water quality and microclimate regulation and control. Nonetheless, the contemporary city, and in particular its coastal areas, is a source of stress and social anxiety, due to the frenetic rhythms that the urban ecosystem undergoes, reacting with ceaseless transformations: the socio-recreational value of the city-sea interface, which influences the liveability, the cultural sphere and the very appearance of coastal cities, plays then an essential role for littoral community. So, it can be said that contemporary coastal society is constantly searching for authenticity along the urban littoral, in response to a sense of loss of culture and identity due to frequent coastal developments (Osbaldiston 2018).

However, the evolution of such communities is constrained by global issues such as climate change, sea-level rise and coastal erosion that may alter the way of living the coast, especially in case of massive urbanization, as emphasised by the *Sustainable Development Goals* of the 2030 Agenda. At the same time, the needs and desires of coastal community members, as well as their rights, have evolved to include issues of spatial and functional equity and environmental justice: for this reason, the resolution of environmental problems cannot be separated from the mitigation of social

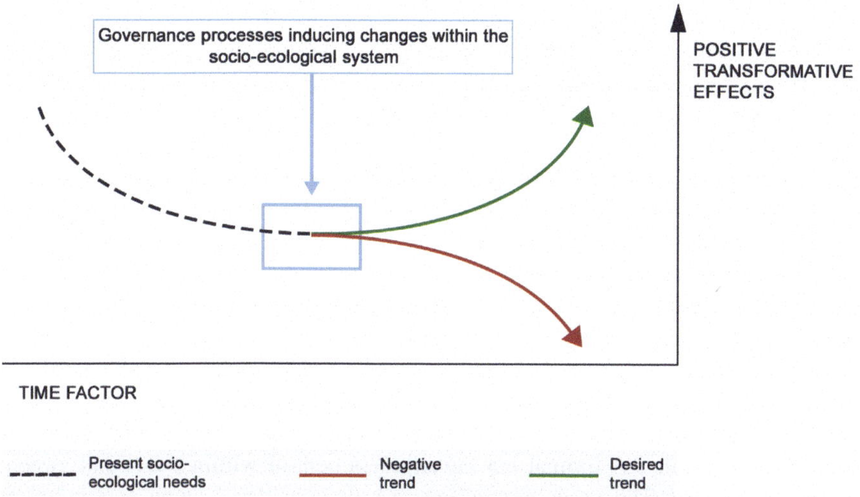

Fig. 3.3 The demands of urban coastal society and littoral environmental needs must converge within governance processes in order to produce positive transformations in the littoral area (elaboration of the author, based on Armitage et al., 2017)

inequalities, because they undermine the liveability of urban public spaces. In accordance with this line of thought, a healthy and safe environment is a fundamental right for coastal communities. Furthermore, equality should be guaranteed to not only between human beings, but also between citizens and urban biodiversity: it is relevant to recognise the weight that the ecological quality of the shore has for the functionality of the city; human beings are therefore a species to be protected as much as their urban-coastal habitat (Lehtinen 2009). From this perspective, it may be reasonable to analyse the different ways users perceive the urban littoral and which functions it is planned for, considering coastal spaces where sociality is now a consolidated habit and places where citizens aim for a redemption of the shoreline for social use (Fig. 3.3).

3.2 Urban Amphibious, Coastal-Urban Forms. How Coastal Society Adapts to the City-Sea Interface

Cities have always risen nearby water bodies: over the centuries, the sea has become an irreplaceable element for communication and economy. Despite the risks and dangers derived from the sea itself, humans have developed defence systems to keep water away from settlements. On the other hand, the perception of the sea and the coast as a place for accessible and equal sociality and recreation is a more recent feature that required new, more strategic, holistic and long-term approaches so that urban coastal society could efficiently and resiliently adapt to the environmental, economic and social changes of urban (Lennon et al. 2014). From a spatial point of view, coastlines are rarely configured as a straight line: this objective fact influences the social relationship with the sea and the organisation of cities. In fact, daily life in constant physical confrontation with the open sea can convey feelings of uneasiness and appear threatening, even in environmental terms. The community has often preferred to gather in compact urban agglomerations that provide a sense of greater protection than the vastness of the body of water and the direct and constant relationship with the open sea.

In this perspective, the notion of *urban amphibious* can fit in: this concept describes the urban contact zone rather than the dividing line between the land and the sea, recalling the capacity of the city and its community to constantly readapting to the two systems, in spatial and functional terms, highlighting how urban coastal society relates to the benefits (but also the criticalities) that the coast puts in place (Worthington 2017). In fact, one of the main peculiarities of the urban amphibious is its location, which is liminal but nevertheless central within the social system of the urban coast. The urban amphibious can be considered as an expression of the city-sea interface: it is a space in which different ways of experiencing the city coast are juxtaposed, from a recreational, commercial, economic but also tourist point of view. In the same way, the flexibility of this area, which recalls the concept of amphibious, have to deal with the environmental evolution of the coastline and the related urbanised zone. It is interesting to investigate the connection between the modification of the coastal interface and the opportunities that the shore offers in terms of functions, as well as the physical factors that influence urban coastal society.

Within the theory of the urban amphibious, it is possible to identify three types of *coastal-urban forms*, namely, configurations that coasts of maritime cities assume in relation to the morphology of the place, the environmental characteristics of the marine space and the specific interventions that society develops with respect to the sea system (Land 2016). The *urban foreshore* represents a limited portion of the coast, related to incoming flows, like tourists and visitors, but also to the social use of the interface by its citizens: it aims to orient users towards the city, directing them towards the services they need. This may include shops, restaurants, commercial spaces, places for refreshment and sociality: historically, the urban foreshore is also a public space that has often played a major political role, similarly to urban areas like public squares that offer a different landscape and planning perspective of the city from the seashore. During the twentieth century, spaces intended for industrial and productive development of coastal areas were in fact gradually converted into pedestrian and cultural areas, with the realisation of museums, parks, public spaces and other places that exploit the proximity with urban water in various ways: perhaps the most representative form of urban foreshore is nowadays represented by urban beaches, whose environmental value and apparent pristine *facies* is somehow preserved with more or less invasive interventions, such as artificial nourishment and

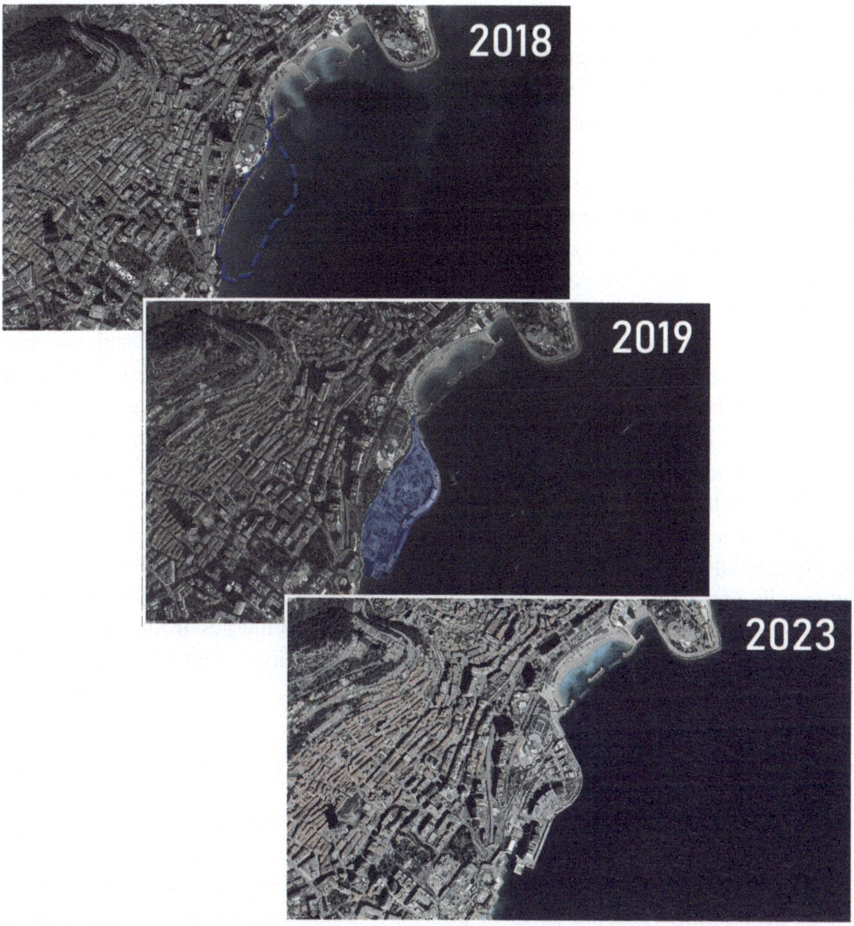

Fig. 3.4 Land reclamation along the urban amphibious of the city of Monaco, in the homonymous principality. The operation is intended to gain about 61,000 square meters of coastal land from the Mediterranean sea (elaboration of the author)

land reclamation,[2] also in relation to coastal erosion and climate impacts (Fig. 3.4). The *urban offshore*, on the other hand, indicates spaces that are more dislocated and peripheral to the urban coastline but equally related to coastal activity, especially from a working, productive and commercial perspective. The relative distance from the neuralgic core of the urban amphibious has led to a relevant coastal economic growth, in relation to the circulation of goods and to logistical features, creating an area that is separated from places for sociality but is characterised by a circumscribed commercial *enclave* (Tagliacozzo 2007). Finally, the *urban estuary* identifies

[2] The process of *land reclamation* refers to the actions aimed at gaining walkable land from the sea, by placing soil or sand in place of portions of the body of water, with consequent ecological-environmental issues (AlQahtany et al. 2022).

particular spots of the urban shore, such as port areas, whose purpose is to guide the outflows of urban coastal society. This portion of the urban amphibious is often linked to variegated socio-cultural expressions: users who frequent this area usually share different languages and dialects and various folkloristic experiences. However, it is not obvious that the ethnic and cultural syncretism within the urban estuary necessarily turns into a convinced coastal cosmopolitanism: usually, it is more likely that this part of the urban amphibious simply coexists with its distinctive social features, even if they can be for sure considered as an enrichment of the coastal interface.

It is evident that each type of urban activity has its own specific spatial dimension that defines the coast, along with an appropriate scale of analysis. As mentioned above for the city-sea interface, the urban amphibious can assume different coastal morphotypes depending on the influences on the natural coastal landscape, the development of the residential fabric and, in general, the urban form: from a geomorphological point of view, the shore can be high and rocky, with cliffs of different heights and eventually beaches at their base, or it can be low with various types of sandy areas, dunes and wetlands. These physical characters influence the width of the three edges of the city-sea interface, especially the maritime-city edge, and relate to the spatial and functional flexibility of the urban amphibious: as it has been shown, it is in fact possible to identify port and non-port littoral zones, peripheral or central areas, commercial spaces and leisure places. However, what emerges from these definitions is the need to delve more deeply into the topic of city-sea interface spaces that are specifically intended for socio-recreational functions, since they have paramount importance for the enjoyment of these areas by the urban coastal society: in particular, *urban blue spaces* demonstrate great complexity, in relation to their conformation as a boundary between land and water, but also in relation to the problem of accessibility, which is often limited only to some beaches or coastal paths.

Moreover, it is well known that users are willing to travel long distances in order to enjoy the urban sea, which reinforces the hypothesis of the crucial relevance of coastal accessibility and the need for proximity maritime recreational spaces (Bell et al. 2020). Understanding users' attitudes and behaviour is very significant for the efficient design of coastal blue spaces: the perceptive impact towards blue-green landscape elements is essential also because it generates interest and attachment to place. Urban blue spaces provide physical and visual access to water and can be described as therapeutic landscapes, a public health resource that promotes psycho-physical well-being and improves social cohesion; for these reasons, urban coastal society usually tends to prefer these areas of the urban amphibious over other natural and semi-natural green areas in the city (Gascon et al. 2017).

It is not a coincidence that the urban amphibious attracts many investors interested in tourism and the amenities of the coast (Fig. 3.5). However, from an ecological point of view, these values are confronted with contemporary problems of coastal erosion and sea level rise, as well as flooding. Finding a solution to these criticalities is by no means obvious or easy: alternatives such as strategic retreat of coastal communities are neither practically nor sociologically feasible, as the extent of the aforementioned issues is difficult to assess, available land is very scarce, and the identity values of

Fig. 3.5 The land reclamation project in Monaco involves the realization of new public green spaces, a promenade and a small leisure harbour. In such interventions, social and environmental demands can easily collide with economic interests (from Wikimedia Commons, licensed under CC BY 3.0)

coastal society are not replicable. Although discussed by academics, urban planners and policy makers, this kind of solutions are not very concrete yet, at least for the moment: to date, there are not many examples of relevant policies or plans that implement such approaches in a definitive and organised manner. While they may be considered efficient adaptation practices because they may put distance between coastal society and risks, they are a cultural hazard with important economic implications: there are countless buildings and spaces threatened by climate change, while relocating the inhabitants of the urban amphibious requires very large sums of money (Gibbs 2016). The vulnerability of this coastal zones is obviously not only measured in terms of housing, but also in terms of infrastructure, environmental contamination, and loss of human lives. It is true that urban coastal society is composed by a broad social spectrum, from the richest to the poorest demographic groups, however, the issue of social equity must be considered, because not everyone has access to the same opportunities, and the wealthiest segments of coastal society undoubtedly have privileged contact with resources, information and services (Siders 2013).

This means that coastal planning must intervene on the functionality of the urban amphibious, protecting its peculiar features and developing a range of services that can satisfy the totality of the urban coastal society, with a view to urban well-being and accessibility to available coastal resources: in addition to the quality of urban green–blue spaces, the urban design should also take into account, therefore, quantitative aspects, like the presence of resting places, the size of pedestrian paths, a satisfying amount of accesses to the urban sea, as these elements are crucial to easily enjoy the city-sea interface, through coastal public spaces that are as inclusive and healthy as possible.

3.3 The Theory of Coastal Commons: Ecosystem Services for Community-Centred Solutions

It is worth asking how to approach technically and sociologically the transformations of the urban amphibious in favour of a less uncertain future development for urban coastal society, considering that cities are not infinitely available to change. Contemporary urban development still seems far from achieving a methodological and normative framework to adaptively counteract the environmental and social problems of cities and urban coastlines: in this sense, green–blue infrastructure can play an important role in reducing land consumption by coordinating the evolution of public coastal space into a coherent landscape and functional network (Adolphe 2022). In particular, the objectives will be the protection and enhancement of the ecological porosity of the urban fabric and the implementation of urban quality and existing open spaces that must coexist with the densification now occurring within the urban system. This could lead to the reutilisation of marginal and underused, even degraded, green–blue public areas, through a process of renaturation and reintegration into the socio-urban system, with attention to environmental risks, strengthening the relationship between the city and open spaces and increasing urban quality standards from a socio-recreational point of view, in favour of the coastal community (Arcidiacono and Ronchi 2021).

In lack of an adequate planning and management response, littoral interfaces nonetheless offer cities the support of *coastal commons*, shared resources at the basis of environmental and ecosystem services on which the city depends and whose importance is crucial for the development of coastal communities (Berkes 2015): these elements are extremely variable according to location and specific local challenges, resulting in similarly articulated governance structures. In fact, the management of the community and environmental components of urban coastlines can only entail a wide range of complexities: the urban amphibious and the coastal society are closely connected by intense human-nature interactions. These interactions themselves are characterised by uncertainty and mutual dependence, needing constant feedback from users and stakeholders (Misiune et al. 2021). The challenges underlying the management of coastal commons are thus equally dependent from both the littoral environment and the urban coastal society: it is necessary to adopt an integrated socio-ecological perspective that mutually and bilaterally relates the human system (within which the coastal society and its socio-economic features for urban well-being are comprehended) and the biophysical system (in which the functional, spatial and environmental characteristics of the urban amphibious can be found). This can have great value for the planning and management of the interface, as from this systemic link it is possible to understand and integrate non-expert knowledge and social needs, governance arrangements based on users' expectations for coastal transformative actions and rules mediating the relationship and interactions between society and its environment.

Communities therefore play a fundamental role in governance, in relation to the desired coastal development but also to the expected egalitarian use of the coastal commons. Furthermore, the concept of scale appears to be a priority for the transformations of the territory of the urban coastal society: however, scale is not limited to the temporal and spatial aspect, but it may also involve the administrative and socio-political sphere. For this reason, a threefold dimensional concept has to be considered: the *geographic dimension* relates to the use of resources on a local or territorial level and to the elements inside or outside the urban-coastal system that influence its use; the *temporal dimension* concerns management development and planning initiatives; finally, the *sociopolitical dimension* is connected to the functional organisation of the coast and its stakeholders (Sowman and Wynberg 2014) (Fig. 3.6).

The coastal society and the city-sea interface possess an adaptive capacity that gives to the socio-ecological system the ability to adapt to external impacts, such as climate change or socio-spatial issues: if the impacts are manageable and do not exceed the adaptive capacity, the urban-community structure will adapt, without changing its identity characteristics; on the other hand, if the impacts are so intense that they exceed the adaptive capacity of the socio-ecological system, the response will no longer be adaptive but transformative, with positive or negative outcomes, depending on the speed of the changes and the overall quality of the relative planning interventions (Armitage et al. 2017).

The concept of the coastal commons is linked to the notion of urban ecosystem services (ES) from which coastal society benefits and which represents a sustainable urban design key to the enjoyment of the city-sea interface. In this context, ecosystem services are considered as an expression of the desires of users, in relation to the biophysical characteristics of the urban coastline, the location of functions and the time necessary to benefit from them, in relation to the overall and peculiar needs of urban coastal society (Villamagna et al. 2013). Urban ecosystems influence the quality of the surrounding environment by regulating rainwater runoff, purifying the air and mitigating microclimates, reducing impending ecological risks; they also

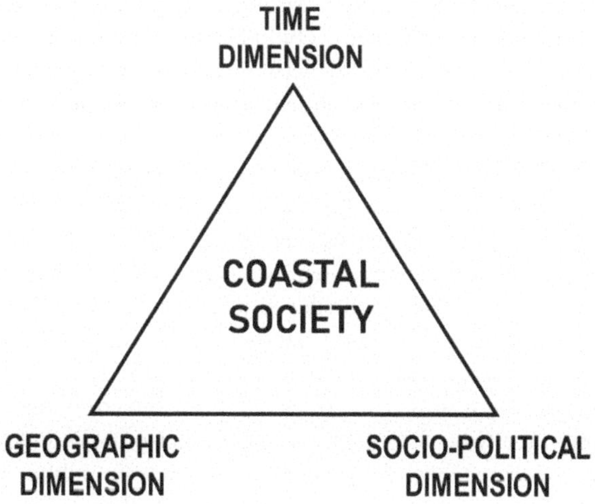

Fig. 3.6 The three dimensions of the socio-ecological perspective within which the coastal commons are framed (elaboration of the author)

increase the aesthetic and perceptive value of the urban system through green areas and open spaces designed for recreation and sociality, offering opportunities for cultural enrichment and preserving local identity and attachment to place through material and immaterial benefits for the well-being of society. This approach is linked to ecological planning and is useful for interpreting the specific needs of contemporary coastal cities, assuming the privileged point of view of citizens, understood as service recipients: the aim is to define a better management of the urban littoral territory, increasing well-being and levelling out social inequalities and orienting urban planning towards environmental implementation, through a more flexible approach in quantitative and qualitative terms (Geneletti et al. 2020).

The demand for coastal ecosystem services and urban benefits is usually estimated on the basis of the needs of the population, focusing urban research on the different levels of accessibility to coastal resources and public spaces. The actual accessibility to public areas that produce costal benefits is evidently differentiated for the various categories of coastal society, generating socioeconomic inequalities and impacts on citizens' well-being (Ronchi 2018). In order to understand the distributive equity of ecosystem services, it is necessary to focus on three main elements: the *supply* locates where services are produced; the *access* indicates by whom and where they are used; the *demand* shows who needs ecosystem services. One of the complexities lies in the fact that the location where ecosystem services are actually provided does not necessarily coincide with the spatial distribution of urban parks or public blue-green areas: in fact, having access to ES is different from having access to the most ecologically efficient areas of the urban system; similarly, the areas where ecosystem services are generated do not always correspond to the places where they can be enjoyed. For this reason, the spatial comparison between ES supply and social demand is crucial to study the actual beneficiaries of ecosystem resources within urban coastal society, although a process of simple spatial mapping might miss whether potential beneficiaries actually succeed in accessing the supply of ecosystem services (Burkhard and Maes 2017). Understanding the dynamic interaction between users' needs and policy makers' strategies can represent an important element in identifying how urban benefits could be equally distributed along the urban amphibious, assessing the relationship between coastal ecological planning and potential accessibility to urban services while taking into account the composition and vulnerabilities of coastal society.

In order to determine the actual accessibility of ecosystem services, it should be identified the flow of ES and the spatial link between the areas in which they are produced and the spaces that benefit from them. Contextual factors that limit the enjoyment of the ecosystem services from the urban coastal society should be then considered: these can include physical barriers, regulatory constraints, environmental limitations, sociocultural habits or practices (Mohr and Dessers 2019); also land ownership can be crucial to assess accessibility to littoral resources, since this factor may prevent some categories of users from enjoying some places along the city-sea interface (Fig. 3.7); there are also other criteria that are more difficult to quantify like, for instance, the perception of safety in public spaces, overcrowding in some places or community cohesion, although they are equally relevant in studying the

Fig. 3.7 Denied accessibility to the sea resource of the urban amphibious at L'Estaque, in the 16th Arrondissement of Marseille. Only users of the nautical club are allowed to use the urban coast (picture by the author)

fruition of the coastal commons provided by the urban amphibious. However, equity does not require that everyone has access to the same types and quantity of ES: on the contrary, the specific needs of each individual or group within the coastal society should be analysed, considering the varied social opportunities and diverse demands in relation to socio-demographic variables (McDermott et al. 2013).

Since each coastal society has its own peculiar relationship with the city-sea interface, the relationships inherent to the socio-ecological system and the coastal commons will vary from place to place: it is therefore not possible to devise fixed solutions for an efficient fruition of urban maritime resources, but rather a different approach should have to be adopted depending on the case, with enough flexibility to identify common factors, examine variables about the urban amphibious and the relationships with the local coastal society that emerge.

3.4 Marine Citizenship: Engaging Urban Coastal Society in Innovative Governance Models

The inseparable link between city-sea interface and urban coastal society opens up to the definition of a *marine citizenship*, that is the awareness of the rights and responsibilities that each citizen should have towards the marine environment that personally experiences, a factor that necessarily influences the ability to manage urban coastal spaces (McKinley and Fletcher 2010). Marine citizenship has its theoretical roots in the conventional notion of citizenship, according to which all users of an urban community should actively contribute to the achievement of collective social, economic and environmental goals (Fletcher and Potts 2007). In general, the definition of citizenship is mainly linked to the public dimension of urban life, considering the value of public commitment that citizens take on through more or less institutional duties and more or less active participation in decision-making processes. At present, this concept also includes the notion of environmental behaviour expected by the society, encouraged at formal and informal levels to reduce ecological impacts: from this point of view, it could be stated that marine citizenship could be an incentive for coastal governance to promote environmental benefits, activating integrated policies (Valencia-Saiz 2005). This concept considers socio-demographic factors and individual and collective values related to the use of the coast and the urban sea: this clearly leads urban planning to identify frameworks based on the inner composition of urban coastal society.

The concept of *marine-ocean literacy* is complementary to the notion of marine citizenship: it aims to encourage a positive behavioural change in the use of the city coastal spaces. In essence, it indicates the understanding that the coastal community has of its influence on the sea and the influence that the sea itself has on the local society (Uyarra and Borja 2016): users with an adequate level of marine-ocean literacy will be able to grasp the basic principles and concepts of the marine ecosystem, actively contributing to spreading knowledge of problems and peculiarities related to the sea and the coast; furthermore, they will be able to take informed and responsible decisions regarding the urban amphibious and its resources (Fig. 3.8).

This means that the notion of marine-ocean literacy is not only an educational supplement, but also a strong incentive to develop true comprehension of the mutual connection between the body of water and the coastal society, for the advantage of the urban-coastal ecosystem. In order to make sure that policy makers and littoral community achieve relevant levels of marine-ocean literacy, it is necessary to investigate the mutual influence and impacts between sea and its users, for example studying ecosystem resources and services, which can be considered an expression of the coastal commons, as stated before. It should be also relevant to translate the human-nature interaction into a positive social driving force from which the coastal-maritime system can benefit: for this to happen, the involvement of urban coastal society is crucial, and it is important to establish effective processes for community listening and methodologies to disseminate coastal knowledge in order to make users as close as possible to marine-ocean literacy (Koutsopoulos and Stel 2021).

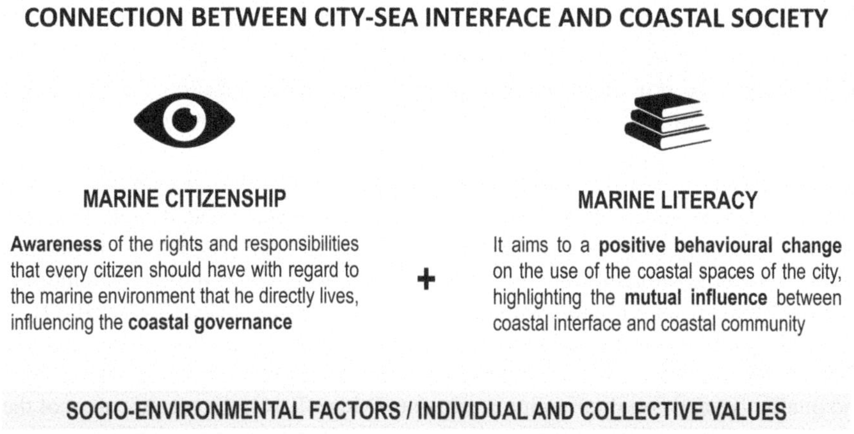

Fig. 3.8 The integration between marine citizenship and marine literacy may improve functional cohesion between the demands of the coastal society and the urban amphibious (elaboration of the author)

It is evident how these concepts are rooted in the spatial organization of urban coastal uses: their integration in maritime planning processes can facilitate coastal design, generating more aware and responsible communities, rethinking open spaces through a diagnostic approach that studies the variables involved in urban coasts and investigating the main features characterising the areas of direct and indirect interaction with the sea, in relation to the progressive littoral urbanisation and the impacts on urban blue spaces (Mishra et al. 2020). On the other hand, the New Urban Agenda promotes the ecological and social functions of public spaces to ensure a progressive improvement of urban quality, focusing on environmental sustainability and ecological design. The functional reorganization of the fragments of the urban amphibious can be implemented in order to efficiently coordinate the existing relationships between the coastal water and the land-sea interface, from a social point of view but also in terms of protection from environmental and hydraulic risks: for this reason, urban planning should promote the connection between the spaces that generate various kinds of ecosystem services with the relative areas where they are actually needed, relating the urban and micro-urban scale with the interventions for risk defence and soliciting community responses to the socio-environmental vulnerabilities of the urban shore (Esmail and Geneletti 2020). These aspects recall crucial decision-making issues that can be framed from a spatial justice perspective: as previously mentioned, this is a complex notion as it is linked to a geographic-spatial sphere, regarding the distribution of resources and users on the reference territory, but also to a social dimension, according to which the community is not homogeneous but it appears to be internally fragmented into subgroups with specific needs and demands. It is therefore relevant that equity becomes a primary objective within decision-making processes, pursuing this value through the active involvement of the population.

Facing these critical issues for the city-sea interface requires design theoretical insights, exploring the concept of planning for public space and coastal green–blue areas, in order to provide support to planners and policy makers. The integrated analysis of the socio-spatial context could provide important knowledge for efficient solutions based on adaptive processes that identify priorities and impacts of coastal interventions. Hence, it is possible to adopt a multidimensional approach that studies urban coasts in relationship with the social perception of spaces, analysing which are the most appropriate coastal uses, the importance that citizens attribute to these areas over time and the influence that the environmental system has on direct and indirect fruition. Strategic anticipation becomes a very useful tool to detect the worst effects of human choices, turning urban planning efforts towards the functional recovery of abandoned and degraded coastal spaces, prone to morphological transformations and qualitative and equalitarian improvements (Mega 2016): the management of the coastal territory should aim for transparent methodologies to reduce the impact of the anthropic component on the city-sea interface, investigating the value of proximity coastal public spaces through sustainable and socially valid functions.

In line with the abovementioned reasoning, the increasing spatial complexity and socio-environmental uncertainty that characterises the contemporary city create an ideal space for participatory innovation and cooperative processes. Such strategies clearly entail a multidisciplinary integration in political-administrative and ecological terms in order to strive for an all-round concept of sustainability: social interactions stimulate the generation of common public visions and they can provide support in identifying emerging challenges, opportunities and threats in advance, if appropriately channelled within strategic anticipation approaches (Garau 2016). In combination with appropriate planning actions aimed at the qualitative improvement of green–blue areas and the creation of new open spaces, community-based approaches to manage urban coastal ecosystem constitute an important opportunity to prevent, control and mitigate the negative impacts of human activities in marine and coastal environments (Beatley 2014), involving in transparent processes all community stakeholders, from political representatives, entrepreneurs and local associations to the common users that compose urban coastal society.

References

Adolphe L (ed) (2022), Integrated Urban Environment Management and Resilience. From the contemporary city to sustainable urbanity. Wiley, Hoboken

AlQahtany AM, Dano UL, Abdalla EME, Mohammed WEM, Abubakar IR, Al-Gehlani WEG, Akbar N, Alshammari MS (2022) Land Reclamation in a Coastal Metropolis of Saudi Arabia: Environmental Sustainability Implications. Water 14(16)

Arcidiacono A, Ronchi S (eds) (2021) Ecosystem Services and Green Infrastructure. Perspectives from Spatial Planning in Italy. Springer, Cham

Armitage D, Charles A, Berkes F (eds) (2017) Governing the Coastal Commons. Communities, Resilience and Transformation. Routledge, London

Banini T (2019) Geografie culturali. FrancoAngeli, Milan

Beatley T (2014) Blue urbanism: the city and the ocean. Islands Press, Washington, DC
Bell S, Mishra HS, Elliott LR, Shellock R, Vassiljev P, Porter M, Sydenham Z, White MP (2020) Urban Blue Acupuncture: A Protocol for Evaluating a Complex Landscape Design Intervention to Improve Health and Wellbeing in a Coastal Community. Sustainability 12(10)
Berkes F (2015) Coasts for People: Interdisciplinary Approaches to Coastal and Marine Resource Management. Routledge, London
Burkhard B, Maes J (eds) (2017) Mapping Ecosystem Services. Pensoft Publishers, Sofia
Clemente M (2013) Identità marittima e rigenerazione urbana per lo sviluppo sostenibile delle città di mare. BDC. Bollettino del Centro Calza Bini 13(1): 181–194
Esmail BA, Geneletti D (2020) Ecosystem Services for Urban Water Security. Concepts and Applications in Sub-Saharan Africa. Springer, Cham
Fletcher S, Potts J (2007) Ocean Citizenship: An Emergent Geographical Concept. Coastal Management 35(4): 511–524
Garau C (2016) Esperienze di pianificazione partecipata. In AA.VV. (eds) Ricerche di Architettura: Nuove Prospettive per l'architettura nella Sardegna del XXI secolo. Gangemi Editore, Rome, p. 10–18
Gascon M, Zijlema W, Vert C, White MP, Nieuwenhuijsen MJ (2017) Outdoor blue spaces, human health and well-being: A systematic review of quantitative studies. International Journal of Hygiene and Environmental Health 220(8): 1207–1221
Geneletti D, Cortinovis C, Zardo L, Esmail BA (2020), Planning for Ecosystem Services in Cities. Springer, Cham
Gibbs MT (2016) Why is coastal retreat so hard to implement? Understanding the political risk of coastal adaptation pathways. Ocean & Coastal Management 130: 107–114
Gillis JR (2012) The Human Shore. Seacoasts in History. The University of Chicago Press, Chicago
Hannigan J (2017) Toward a Sociology of Oceans. Canadian Review of Sociology 54 (1): 8–27
Koutsopoulos KC, Stel JH (eds) (2021) Ocean Literacy. Understanding the Ocean. Springer, Cham
Land I (2016) Doing Urban History in the Coastal Zone. In Beaven B, Bell K, James R (eds) Port Towns and Urban Cultures. Palgrave Macmillan, London, p. 265–281
Lehtinen AA (2009) Environmental Justice. International Encyclopedia of Human Geography: 535–539
Lennon M, Scott M, O'Neill E (2014) Urban Design and Adapting to Flood Risk: The Role of Green Infrastructure. Journal of Urban Design 19(5): 745–758
McDermott M, Mahanty S, Schreckenberg K (2013) Examining equity: A multidimensional framework for assessing equity in payments for ecosystem services. Environmental Science & Policy 33: 416–427
McKinley E, Fletcher S (2010) Individual Responsibility for the Oceans? An Evaluation of Marine Citizenship by UK Marine Practitioners. Ocean and Coastal Management 53(7): 379–384
Mega VP (2016) Conscious Coastal Cities. Sustainability, Blue Green Growth, and The Politics of Imagination. Springer, Cham
Mishra HS, Bell S, Vassiljev P, Kuhlmann F, Niin G, Grelli J (2020) The development of a tool for assessing the environmental qualities of urban blue spaces. Urban Forestry & Urban Greening 49
Misiune I, Depellegrin D, Vigl LE (2021) Human-Nature Interactions. Exploring Nature's Values Across Landscapes. Springer, Cham
Mohr BJ, Dessers E (eds) (2019) Designing Integrated Care Ecosystems. A socio-technical perspective. Springer, Cham
Osbaldiston N (2018) Towards a Sociology of the Coast. Palgrave Macmillan, London
Ronchi S (2018) Ecosystem Services for Spatial Planning. Innovative Approaches and Challenges for Practical Applications. Springer, Cham
Schwermer H, Barz F, Zablotski Y (2019) A Literature Review on Stakeholder Participation in Coastal and Marine Fisheries. In Jungblut S, Liebich V, Bode-Dalby M (eds) YOUMARES 9 - The Oceans: Our Research, Our Future. Springer, Cham, p. 21–43

Siders A (2013), Managed Coastal Retreat: A Handbook of Tools, Case Studies, and Lessons Learned. Columbia Center for Climate Change Law, New York
Smith TF, Thomsen DC (2008) Understanding Vulnerabilities in Transitional Coastal Communities. In Wallendorf L, Ewing L, Jones C, Jaffe B (eds) Proceedings of Solutions to Coastal Disasters, Hawaii. American Society of Civil Engineers, Reston, p. 980–989
Sowman M, Wynberg R (eds) (2014) Governance for Justice and Environmental Sustainability. Lessons across Natural Resource Sectors in Sub-Saharan Africa. Routledge, London
Tagliacozzo E (2007) An Urban Ocean: Notes on the Historical Evolution of Coastal Cities in Greater Southeast Asia. Journal of Urban History 33(6): 911–932
Thomsen DC, Smith TF, Carter RW, Mayes G (2009) Defining Community: Understanding the Meaning of 'the Community' in Coastal Zone Management. Journal of Coastal Research 56: 1316–1319
Timur UP (2014) Urban waterfront regenerations. In Ozyavuz M (ed) Advances in Landscape Architecture. InTechOpen, London, p. 169–206
Uyarra MC, Borja A (2016) Ocean Literacy. A New Socio-Ecological Concept for a Sustainable Use of the Seas. Marine Pollution Bulletin 104: 1–2
Valencia-Saiz A (2005) Globalisation, cosmopolitanism and ecological citizenship. Environmental Politics 14(2): 163–178
Villamagna AM, Angermeier PL, Bennett EM (2013) Capacity, pressure, demand, and flow: a conceptual framework for analyzing ecosystem service provision and delivery. Ecological Complexity 15: 114–121
Worthington D (ed) (2017) The New Coastal History. Cultural and Environmental Perspectives from Scotland and Beyond. Palgrave Macmillan, London

Chapter 4
The Effects of Climate Change on Contemporary Urban Coastlines

Abstract Climate has a significant impact on human well-being, particularly in coastal areas where meteorological elements affect communities. Understanding the influence of the atmospheric environment on human comfort is crucial for planning the city-sea interface, considering factors such as leisure activities, tourist industry, energy production, land use management and ecological issues. Integrated climate risk management is becoming increasingly important in this context. However, awareness of climate risks varies between stakeholders, with non-experts often underestimating potential impacts. Bridging this gap is essential for effective participatory coastal planning, as lack of climate awareness is a major challenge to effective management of maritime risks.

Keywords Climate change · Coastal risks · Environmental planning · Coastal adaptation · Climate awareness · Risk perception

4.1 Risks and Vulnerabilities of the City-Sea Interface

Urban planning assesses the presence of risk as an element that can negatively affect the quality of life of citizens, in different areas related to the fruition and structure of the territory. It is possible to distinguish some categories of risk: natural risks are linked to disastrous phenomena such as floods, coastal storms, earthquakes or volcanic eruptions; technological-environmental risks are generated as a response to the possible damage that human action can bring to the surrounding ecosystem, such as explosions, fires or the greenhouse effect; finally, social risks involve the more or less realistic perception of public space, in relation to the degree of frequentation and spatial degradation (Pine 2014). On the basis of these definitions, it is possible to identify climate risk as the probability that a certain critical event will occur in a specific place due to direct or indirect causes, causing more or less loss of lives, damage to the landscape-cultural heritage and the surrounding environmental ecosystem. Its extent can be determined through the product of hazard, vulnerability and exposed value. Hazard indicates the potential occurrence of a negative

I. Pistone, *Urban Coasts in Socio-ecological Transition*,
https://doi.org/10.1007/978-3-031-70783-4_4

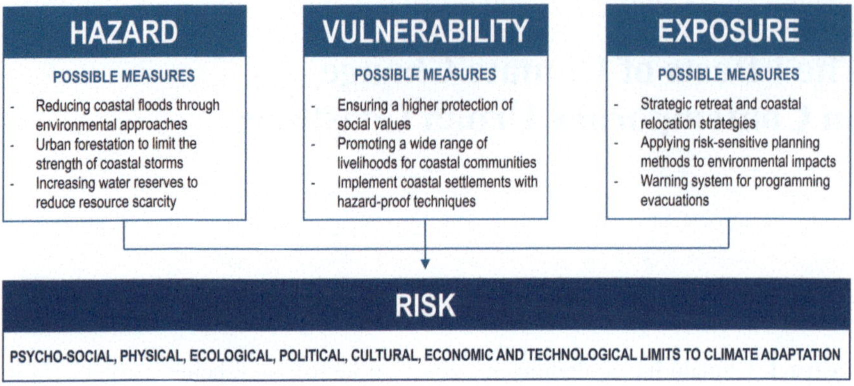

Fig. 4.1 The climate risk function. It is possible to act from a physical and social point of view on the various factors underlying the risk; however, it is necessary to consider limits and criticalities of different nature (elaboration of the author)

event or impact of anthropic or physical-natural kind that brings damage to users, building, facilities and the available ecosystem resources; vulnerability is instead the systemic or territorial capacity to endure the negative effects of climate impacts; finally, exposed value refers to the elements potentially influenced by a certain risk, considering the natural or built environment and the population among these (Field and Barros 2014) (Fig. 4.1).

Current studies show that climate is rapidly changing and will continue on this trend, altering both societies and the environment: there is much scientific evidence of the worsening of weather and water phenomena in the near future (Timmerman 2021; Allen et al. 2018; Blöschl et al. 2017). The environmental crisis becomes more and more relevant in terms of academic research, urban planning practice and contemporary public space management, especially in coastal areas, where most cities are concentrated and thus suffer the effects of climate change, progressively more intense. It is estimated that the environmental impacts will be increasingly evident and catastrophic, confirming the risk that coastal settlements run more than other anthropized areas of the planet: on average, there has been a worrying increase in global temperature, fluctuating between minimum peaks of 0.8 °C and maximum peaks of 1.2 °C, in comparison with the pre-industrial period; this situation will tend to worsen up to an increase of 1.5 °C if adequate preventive measures will not be taken between 2030 and 2050 (Fig. 4.2).

The effects of global warming have had major negative implications on the intensity and frequency of extreme climate events, causing an exacerbation of heat waves, floods and coastal storm, and long-term changes, such as the progressive worsening of sea-level rise (Masson-Delmotte and Zhai 2021). These conditions clearly have tangible consequences on the socio-environmental systems at the base of the communities living in the city-sea interface, imposing sustainable approaches to coastal planning to cope with the rapidly evolving impacts of global warming: its

Fig. 4.2 Heavy swell in Barcelona, triggered by storm Gloria that hit the Spanish coast in 2020 (from Wikimedia Commons, licensed under CC BY-SA 3.0)

effects are increased by the heavy urbanisation of coastal settlements, the thermo-hydraulic properties of buildings and the scarcity of vegetated areas (Gaglione 2022). Urban coastal areas are in fact characterised by an intrinsic complexity and dynamic morphology, due to their nature as both border and transition zones between the city and water system, which exposes them to multiple risks associated with the dangers and increasing effects of climate change.

Climate has a profound influence on human well-being and health: meteorological elements can indeed act individually or in combination to produce effects on communities. Assessing the influence of the atmospheric environment on human comfort becomes a topic of great interest in the context of life along the coastal interface, as a function of users' leisure, tourism, energy industry, urban planning and architecture, and quality of coastal waters. For these reasons, integrated climate risk management acquires more and more relevance within coastal spatial planning: vulnerability to climate change is interpreted in a proactive way, focusing design and management efforts not only towards preventive actions for catastrophic event, but also in the stimulation of social awareness of risks and adaptive responses by the site itself; identifying and assessing the aforementioned parameters of hazard, vulnerability and exposure become fundamental for an efficient climate risk management (Pörtner and Roberts 2022).

Sea-level rise represents one of the main critical issues for the city-sea interface, also in its insular areas: in the last century, the sea level has risen by as much as 25 cm, increasing exponentially in recent years (Moatti and Thiebault 2016). Flooding constitutes an additional element of risk for coastal use, as it involves the overflowing of the regular banks of a water body or the accumulation of water in areas where usually it should not be found. In urban coastal areas, floods can occur either due to natural causes, such as intense flooding, or due to anthropogenic causes, such as a malfunctioning sewage disposal system. It is possible for coastal floods to occur suddenly (*event-based hazards*), as in the case of unexpected storm surges, or gradually, as in the case of progressive sea-level rise that increases over a long-term time. It is also not uncommon for event-based hazards and gradual hazards to occur at the same time, increasing their specific effects and intensifying impacts on coastal settlements (Rholi and Li 2021). In this case, it is possible to refer to the so-called *extreme sea level events*: such phenomena can occur during severe storms and cause heavy floods if protection works are lacking or inadequate. It is evident how this is harmful to public life along the city-sea interface, as the sea level rise can easily reach 10 cm, a value that can be increased depending on the strength of the tides, wind-related wave motion and possible cloudbursts, as well as the morphology of the coast itself (Bevasqua et al. 2019): without efficient risk reduction operations, damage in Europe is estimated at almost one trillion euros within the first century of the new millennium (Ciscar et al. 2018).

One of the main consequences of these events is the coastal erosion that affects contemporary coastal cities, namely the natural or induced processes that alter coastal morphology, leading to a gradual loss of emerged land: this is measured in relation to the time duration of the event and to the dynamic variation of the sedimentary littoral, intended as the transition area between the land and the sea. According to the data provided by the EMODnet platform,[1] this phenomenon is quite stabile, even though some locations present critical peaks in negative morphological variation, especially in Italy, Spain and along the Balkan peninsula; it is however possible to state that as much as 28% of European coasts is progressively eroding, in relationship with the data of the European project EUROSION, one of the very few pan-European analysis assessing to date the effects of coastal erosion on a continental scale: about 15,000 km of coastline affected by environmental impacts are actively receding, even where risk defence works are actually located, while only 4700 km have been effectively stabilised (Doody et al. 2004).

Therefore, all European coastal states are affected by coastal erosion, coastal flooding and climatic risk to some extent, although in the Euro-Mediterranean area, these critical issues appear even more complex: in fact, the historical and cultural heritage has a priceless value that cannot be in any case replicated. The Euro-Mediterranean city-sea interface is characterised by the presence of approximately fifty UNESCO sites: their relevance is connected to the economic, social and identity sphere of urban coastal society (Reimann et al. 2018). Thus, the close relationship between socio-economic development models and environmental implications emerges: not only the anthropic action produces climatic alterations increasing risks, but the usability of the littoral public space is clearly limited by phenomena such as floods and storm surges. Moreover, there have been numerous climate catastrophic

[1] The *European Marine Observation and Data Network* (EMODnet) is a long-term initiative to collect and disseminate multidisciplinary data on marine environments funded by the European Commission (https://emodnet.ec.europa.eu/en).

events that have affected coastal cities in recent years, in various parts of the Euro-Mediterranean basin (Benito et al. 2015), while sea-level rise future projections report an impressive increase from 20 to 57 cm between 2050 and 2100 (Vecchio et al. 2019).

The consequences of climate change require contemporary planners and decision makers to devote attention and design efforts for realizing adaptive and resilient public spaces that can withstand environmental impacts from the sea: for this to happen, the spatial distribution of hazards, uncertainties, vulnerabilities and communities at risk along the city-sea interface must be assessed. Indeed, climate risk is not evenly distributed across the territory, just as some branches of urban coastal society may be more sensitive than others to environmental impacts. A spatial mapping of the elements affected by climate risk can therefore be very relevant for planning and management in the present and in the future. Also, the social awareness is linked to this reasoning. Awakening the consciousness of urban coastal society has crucial relations with environmental planning and design: the communities are called upon to play a more proactive role in governance processes, through open dialogue, social listening and collaboration among stakeholders. The ultimate goal is giving to the city-sea interface the ability to quickly regain its equilibrium after eventual disastrous events generated by climate change, always guaranteeing a high quality of public space for the coastal society (Tubridy et al. 2021).

4.2 Environmental Planning and Adaptation for Coastal Climate Neutrality

Climate protection comprehends policies and planning actions aimed at progressively reducing the impact of climate change on natural and human systems, while limiting environmental externalities that may favour weather changes in the short and long term: the joint and multi-scale approach of coastal adaptation and climate risk mitigation is proposed as an application of the principles of environmental planning, from both a top-down and bottom-up perspective (Musco and Patassini 2012).

At the EU and national level, policies and directives for reducing greenhouse gas emissions in the atmosphere and the sustainable transformation of the contemporary city and urban coastline are becoming increasingly widespread. At the root of this long process, it is possible to place the Rio Agreement, signed in 1992 following the United Nations Framework Convention on Climate Change (UNFCCC),[2] recognising the importance of human action in climate change, in terms of damage caused but also in relation of future potential solutions. The Goal 13 of the 2030 Agenda,

[2] The UNFCCC has promoted several major conferences among its approximately 200 member states: one of the most important outcomes of the Conferences of the Parties (COPs) are the Kyoto Protocol (1997), which legally binds the member industrialised countries to reduce greenhouse gas emissions since 2005, and the Paris Agreement (2015), which tries to limit global warming to no more than 2 °C above pre-industrial levels. The next COP will be held in November 2024.

endorsed by the United Nations in 2015, highlights the importance of defining appropriate climate actions, also in relation to the record-breaking temperature increase registered in 2019, considered the second hottest year of all time, in the hottest decade ever (2010–2019); despite the fact that emissions of climate-changing gas have fallen by 6% following the lockdowns imposed in 2020 all over the world due to Covid-19, it is still imperative to take action against climate change, as underlined by the Goal 11 of the 2030 Agenda which aims to reduce pollution produced by individuals in the cities, pursuing an integrated and participatory development for settlements, based on the integration of the environmental, social and economic dimensions of sustainable development (Fig. 4.3).

From the perspective of climate adaptation, through the *EU Adaptation Strategy* (2013), the European Union aims to encourage member states to have their own national strategy, integrating a resilient approach within sectoral policies, and to ensure decision-making processes linked to climate adaptation theories; in 2021, the strategy was updated, defining a time horizon up to 2050 and promoting systemic climate risk management at every level of governance, in order to respond adaptively and efficiently to impacts.

In the context of coastal zones, the European Community has been expressing its views on the protection of shorelines and the settlements along them since 1976 with the Barcelona Convention, later updated in 1995. However, at the beginning of the new millennium, the need for efficient maritime-coastal planning led to the drafting of the EU *Integrated Coastal Zone Management Directive* (ICZM) (2002), aimed at an integrated, dynamic and continuous management for a conscious and sustainable use of littoral areas. In 2013, the EU *Maritime Spatial Planning* was approved: it is one of the main tools for the implementation of ICZM, due to the strong environmental, social and economic interactions between land and sea along the coastal interface.

The *EU Flood Directive* (2007) is also relevant for studying the management of the urban amphibious in critical climate scenarios: it aims to assess and contain flood risks for member states, while integrating adaptive measures against coastal storms. The EU *Marine Strategy Framework Directive* of 2008 proposes strategic approaches to face human-induced changes to littorals. During this post-pandemic period, it makes sense to mention the financial support of the *NextGenerationEU* Recovery Plan, which the European Commission launched in 2021 to relieve the current critical socio-economic and environmental condition by allocating more than 800 billions, pursuing ecological resilience in communities.

On the basis of this community framework, it is possible to define the main characteristics of ecological planning applied to the coastal interface context for climate risk defence, facing this issue in a preventive manner or through lighter or heavier solutions, depending on the specific case. The term mitigation identifies interventions and preventive measures that planners can take to reduce climate-changing gas emissions, thus acting on the causes of climate change rather than its effects, limiting land consumption and the impacts caused by sectors that are most responsible for pollution, such as the logistics and productive fields: historically, climate change has been caused mainly by intensive land use, exceeding by 25% the impacts generated from

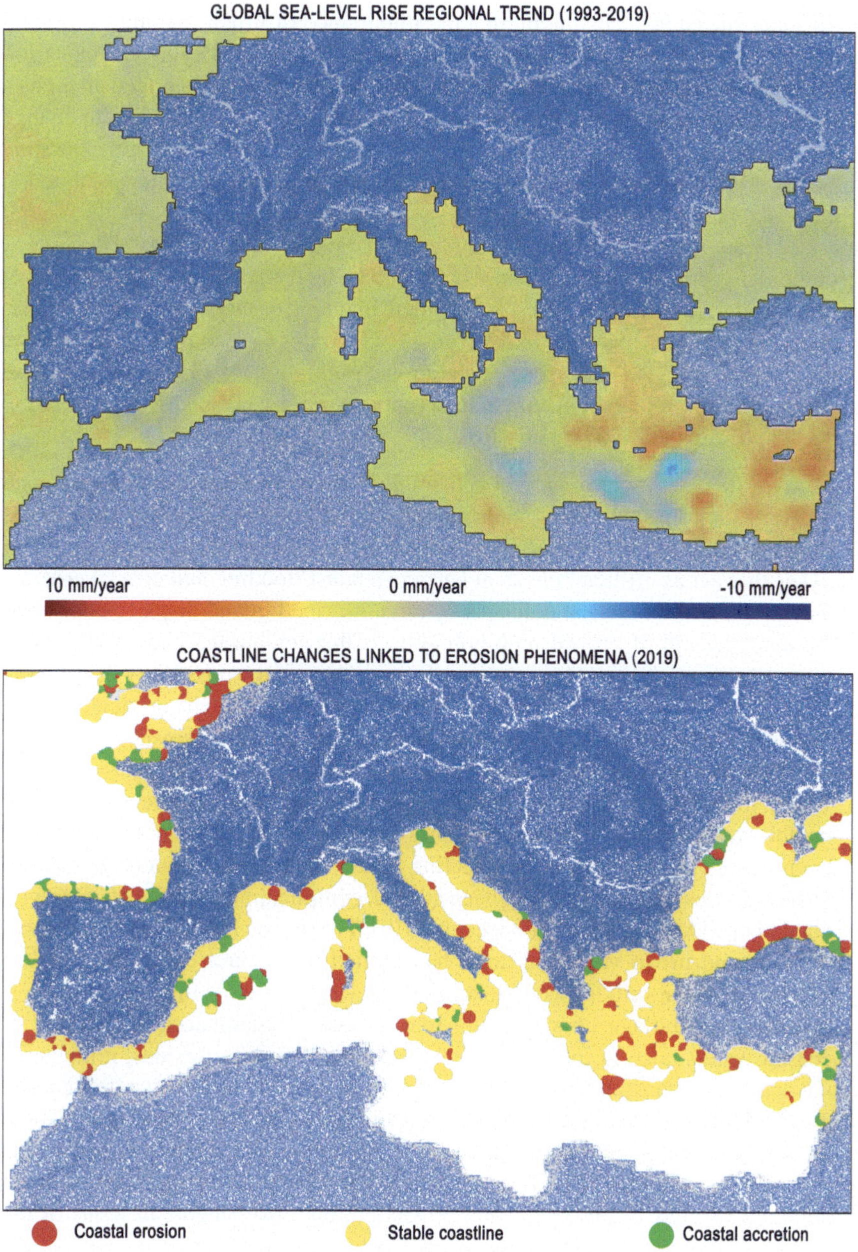

Fig. 4.3 Above, sea-level rise mostly affects the eastern side of the Mediterranean sea. Below, coastal changes in the Euro-Mediterranean basin, within which there are some cases of serious criticality (elaboration of the author, based on EMODnet data)

fossil fuels emissions. This highlights the importance of urban planning and valid policy making to limit the increasing environmental damage.[3] Moreover, adaptation strategies are described as preventive or reactive actions to climate effects that gives to the coastal interface a sufficient degree of resilience to live with climate change in the short and long term, minimising negative impacts and anticipating possible damages (Manigrasso 2013). Currently, there is a wide range of strategies aimed at climate adaptation of the urban amphibious. It is possible to intervene on the coastal interface in a purely technological manner (as in the case of fixed coastal defences), through management interventions, which modify land uses according to the level of risk, through political decisions that regulate planning criteria for the urban shore; besides, behavioural measures may be adopted in order to reconfigure the fruition of the city-sea interface, also from a recreational point of view: on the basis of these premises, it is possible to define three intervention strategies listed below, in relation to the adaptive response of the coastline.

– *Defence strategies* seek to protect vulnerable elements of coastal settlements, such as users, facilities, buildings and socio-economic activities, through appropriate tools. This includes the installation of fixed defensive elements, such as reefs or dykes that act as artificial physical barriers against flooding and coastal erosion, the realignment of coastal defences by circumscribed inland displacements, and the management of coastal morphology to better contain the causes of climate impacts: among the most innovative measures, there is the possibility of reducing wave intensity by moving the coastline further in the sea from its original location through beach nourishment, building offshore barriers or even installing specific maritime devices useful to generate renewable energy (Pecher and Kofoed 2017). Within this category, it is also possible to group solutions with a greater degree of flexibility, such as the realization of dunes or urban vegetation works: as a result of the progressive cementification affecting urban settlements, the delicate balance of coastal ecosystems has in fact been altered, complicating the lives of users with phenomena like heat islands (Wu et al. 2019). The drastic reduction of permeable surfaces and urban vegetated areas has limited the natural capacity to absorb exceeding water, critically overloading the sewage systems: the consequence is therefore the spillage of wastewater into the sea, contaminating the quality of coastal waters, also in relationship with bathing activities. Besides, it increases the risk of hydraulic and hydrogeological criticalities that frequently affect coastal areas (Mendonça De Carvalho and Szlafsztein 2019). So, the presence of resilient open spaces and a valid connection to the green–blue infrastructure at multiple planning levels within the city-sea interface can have positive implications from the point of view of environmental protection, but also benefits in terms of the psycho-physical well-being of the urban coastal society, which could enjoy much higher quality coastal public spaces thanks to the influence of urban vegetation nearby the sea (Massoni et al. 2018). In addition to this, various tools can be

[3] The European Union has funded the CCTAME (Climate Change—Terrestrial Adaptation and Mitigation in Europe) project to enhance land use and policy by devising different scenarios of climate impacts from a preventive perspective.

developed to prevent the maritime water infiltrations, thus reactively reducing the risk from climate change. Such expedients aim to defend the urban coastline from environmental hazards by limiting land-use changes as much as possible.

- The *adaptation strategies* are intended for the occupation of areas of the coastal interface with high sensitivity to climate risk, modifying their fruition in relation to potential scenarios of strong sea-level rise and consequent flooding in order to amplify the capacity of the urban littoral to respond to negative impacts. This kind of approach is aimed at a more flexible and proactive risk management, adapting spatial functions and human activities to environmental conditions with a view to damage reduction: the possibility of potential sea hazards can be therefore acceptable as a function of normative guidelines for climate resilience of new and existing buildings, for creating shelter points in case of critical events, or for choosing a type coastal vegetation (as well as possible coastal plantations) that better resists the proximity of sea water and potential flooding (Breil et al. 2007). Adaptative interventions also include sustainable and green solutions: these can be categorised as *Ecosystem-based Approaches*, integrated strategies aimed at increasing settlement resilience in order to achieve a wide range of environmental, economic, social and cultural benefits: within these approaches, solutions such as nature-based solutions, green–blue infrastructures, sustainable urban drainage systems and water sensitive urban design can be included (Bergier and Kowalewska 2019). The growing environmental focus on sustainability shifted the core of urban design concepts and practices in the planning of natural areas, public spaces and green zones within the urban system. Sea risk defence strategies that rely only on grey infrastructure elements, such as dykes and artificial maritime shores, have demonstrated to be insufficient in dealing with large-scale climate events, like mega-floods with return ranges of more than one hundred years: there is therefore a need for an effective integration between the advantages offered by technological solutions and the many benefits that innovative green interventions bring to the protection of the most vulnerable areas along the city-sea interface (Nakamura 2022). However, in 2021 the European Commission identified a general slowdown in the adoption of coastal adaptation strategies due to a lack of knowledge about climate risks, weak decision-making support in the planning stages, but also a weak monitoring and feedback system of the ongoing projects, which require a very long time to be implemented: moreover, at a community level, it is not common to take into account climate impacts also outside the European framework, which could provide management and planning insights, as well as interesting best practices. In addition to these approaches, there are also strategies aimed at enhancing the resilience of the city-sea interface: to achieve this goal, it is possible to intervene on the systemic vulnerability through structural changes to the facilities and buildings that are exposed to impacts, but also through decision-making processes for improving the environmental adaptability of the socio-economic component of coastal areas. In this sense, the participation of the stakeholders of urban coastal society becomes fundamental because it may lead to a shared vision of the potential effects of climate change on the urban amphibious and to more efficient plans and actions for adapting the littoral feature to the contemporary climate status.

- *Relocation strategies* propose the planned retreat of coastal society from areas of the city-sea interface affected by higher environmental risk to new areas that could be located inland or in other less critical points of the urban coastline, in order to escape potential damage and negative impacts (Evans et al. 2004). This solution does not involve specific planning efforts to protect areas at risk of flooding or coastal erosion, also because of the high costs for realising an effective urban shoreline defence. The extreme scenario foresees the total abandonment of the coastal area: this kind of strategy tries to relocate economic development to other territories, while preserving as far as possible the tourist value of the original coastal area after sea level rise and flooding events. However, as it has already been discussed before, it is not always possible to replicate the peculiar identity values of the city-sea interface, also considering how low is the available land resource today (Fig. 4.4).

It should be emphasised that today climate planning is not focused on a genuine holistic perspective, since it is still considered much more as technical-administrative prerogative of specific mono-disciplinary bodies, generally related to civil protection and energy management: much remains to be studied in this field, especially with

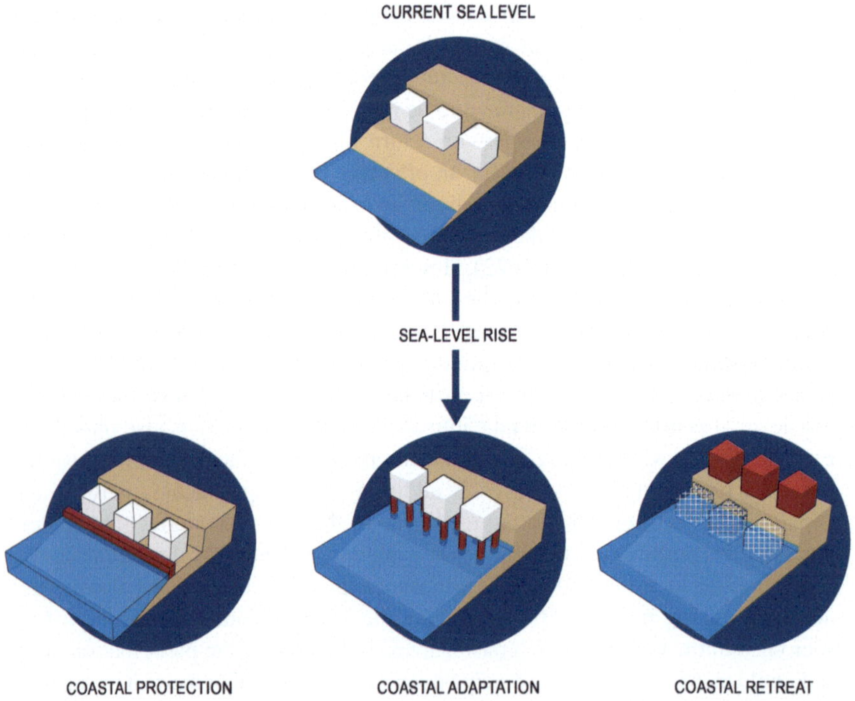

Fig. 4.4 Possible approaches to face sea-level rise and its related effects (elaboration of the author)

regard to the weather sensitive and vulnerable spaces along the city-sea interface. At the local level, it is possible to identify in different global contexts the Climate Plan tool, elaborated in various ways depending on the geographical and environmental context: generally, the plan defines a strategic framing document, which identifies long-term objectives on the basis of socio-economic and environmental characteristics, analysing socio-economic and environmental factors, the operational fields and the involvement of both public bodies and common users with specific participatory methodologies; an emission inventory is then elaborated in order to assess the economic elements and sectors to change for climate mitigation purposes; at the operational level, the appropriate actions are selected, identifying potential impacts for short and medium-long term scenarios and indicating areas affected by hydraulic risk, difficult runoff, urban heat island formation and coastal erosion. However, still little steps have been made in Europe towards the integration of climate policies within urban planning and decision-making processes for the territorial governance of climate risk, especially in relation to urban coastal spaces (Musco and Fregolent 2014). In fact, climate protection plans and directives are part of a multi-level framework in terms of scientific fields but also application scale, from the territorial dimension to the local government. At the municipal scale, it is not always possible to manage the complexity of coastal climate proof planning, even though it is fundamental to comprehend local public spaces along the coastal interface in this kind of approach. Even if decisions about short-term actions or energy standards for buildings manage to fit into small-scale governance quite easily, there is still need for an integrated framework to coordinate territorial and national policies with local authorities for long-term interventions regarding coastal land uses and accessibility (Musco and Van Staden 2009).

4.3 Perception and Awareness of Urban Coastal Society on Climate Change

The consciousness of climate risk along the city-sea interface may differ from user to user for many reasons: first of all, the perception experienced by non-expert stakeholders living in areas at risk could vary from the knowledge possessed by expert individuals, such as planners and decision-makers. For this reason, users who are most affected by maritime risks tend to underestimate the potential impacts of catastrophic environmental events: within participatory approaches to coastal planning, the lack of climate awareness in the communities settled nearby the sea represents a major challenge for an efficient management of maritime risks (Chauvin 2014).

When climate change started to become a very relevant topic among scholars in the early part of the 1980s, the social perception of this phenomenon was on average low levels. Public concern, however, rose rapidly over the next twenty years, especially in heavily industrialised countries, as a consequence of growing scientific evidence, progressive media coverage and frequent political debates. In general terms, the international awareness of the urgency for climate change mitigation has been reached in the first decade of the new millennium: yet, from the perspective of the average users, knowledge about the effects of human activities on the current environmental situation has not increased particularly over time (Baiardi and Morana 2021). Indeed, there is a general lack of awareness from urban coastal society regarding these issues: this undermines the efficiency of planning in supporting design efforts for littoral adaptation. Moreover, the community usually resists adapting to new practices and uses considered conflicting with common habits, worrying more about coastal risk management policies rather than the environmental risk itself (Krien and Michel-Guillou 2014).

In this sense, the theory of *social representation*[4] can be useful to understand coastal society users' perceptions of environmental risk, in order to calibrate behaviours to be encouraged or legitimised in policy making and urban planning, overcoming the gap between objective and subjective risk awareness. The social representation indicates the reorganisation of socially constructed opinions on a topic, in this case coastal risk. The development of a common knowledge enables individuals to cope with its complexity but it is also a driving force for certain actions on the part of coastal society: the understanding of these elements sheds light on the meaning that potential environmental impacts have for the citizens of the city-sea interface. In case of floods and coastal erosion, individual opinions of users are not very coherent, because different social groups give different meanings to the main elements associated with climate issues. Besides, the critical situation of coastal hazards for the city-sea interface justifies the psycho-social approach to understand how society identify and reacts to risks from the sea (Lemée et al. 2019) (Fig. 4.5).

For urban planners and decision makers, it is therefore complex to precisely communicate the imminence and extent of risks because of the differentiated perception of the coastal society about environmental threats, considering influencing factors like the socio-cultural background, social conventions and habits (Lieske et al. 2014). Risk perception is also connected to how long users have lived in the same place: exchanging opinions and experiences between recent and long-time inhabitants can therefore induce shared attitudes towards negative events such as sea-level rise or coastal floods, consequently influencing social representation at the local level (Fig. 4.6).

In this perspective, the concept of urban identity is recalled. In urban studies, identity can be described on the basis of the special characteristics of a place and its value attributed by individuals. It is a way of representing the natural, socio-cultural and artificial components of the city; its usefulness lies in defining the actual appearance of a space, in relation to its social representation, recognising the physical, functional and social elements that are common to all users (Luo et al. 2023). Urban identity can be analysed according to three basic components: the cognitive aspect defines

[4] *Social representation* refers to the transmission of common knowledge within certain social groups through which individuals can organise and interpret information from their own context: this theory is useful in spatial management even if representations may vary from one social group to another (Van Lange et al. 2011).

Fig. 4.5 Rally in New York, as part of the Fridays For Future initiative to raise awareness of climate change (ph. F. Davis, from Wikimedia Commons, licensed under CC BY 2.0)

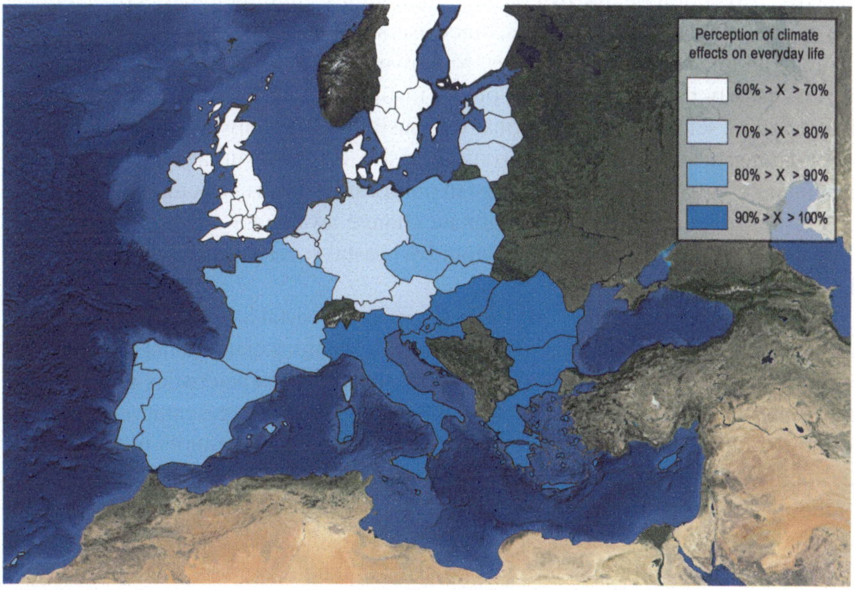

Fig. 4.6 Survey aimed at assessing the perception of impacts that climate change has on the lives of European citizens (period 2019–2020): higher values are recorded in the southern area of the continent (elaboration of the author, based on European Investment Bank data)

the attachment of an individual to an urban system as its member; the affective aspect emotionally connects the individual to the urban space; the evaluative aspect is linked to the urban quality that the user gives to a place. The social representations of the community influence the relationships between these three domains of urban identity, so expert researchers should relate to the climate risk assessment directly produced by non-expert users in order to clearly communicate the effects of environmental impacts to the coastal society (Belanche et al. 2017).

However, the perception of environmental risks related to climate change presupposes a certain degree of awareness of their existence and the potential damage they can generate, although this does not prevent citizens from choosing to enjoy vulnerable spaces such as urban coastal areas. The littoral community is also influenced by less rational components in the perception of risk, such as the emotional aspect and feelings: public spaces are in fact places widely frequented by citizens and can potentially assume an important role in improving urban life, affecting the community's sense of place (Bollini 2018). The well-being derived from the fruition of the city-sea interface is indeed a parameter for assessing the quality of coastal settlements and can be linked to the evaluative aspect of urban identity, but also to the affective one: this reduces the possibility that urban coastal society decide to move away from the areas that are threatened by environmental criticalities, as they appreciate the proximity to the sea and their social relationships are also based on the peculiarities of the urban coastal space. It is evident that the place attachment may interfere with the objective perception of risk by coastal communities, limiting the effective application of coastal adaptation solutions, since users may feel reticent towards particular planning and management strategies. Moreover, as previously mentioned, ordinary citizens usually trust quite little in public institutions regarding topics like climate change; at the same time, institutional responsibility is not always integrated within actions aimed at improving the environmental situation of the urban amphibious. This is an interesting fact, since a solid trust in public authorities can actually corroborate the existing relationship between climate change awareness and the development of adaptive and positive social behaviour by coastal society (Schleyer-Lindenmann et al. 2022).

It could therefore be stated that the perception of coastal hazards within the urban coastal society constitutes an important element within social practices, although it varies according to the spatial context, environmental characteristics and personal experiences of the users. Understanding the value of social representations proves to be an important cornerstone for analysing individual and community perceptions of environmental hazards: considering that risk management policies are often invasive for the population and generally not well received, the personal involvement of coastal community users appears necessary for an efficient ecological planning of the city-sea interface (Michel-Guillou and Meur-Ferec 2017). Public opinion should be considered valuable by policy makers: if they manage to decode the socio-environmental variables at the base of the social representation of climate risk, they may actually strengthen environmental mitigation policies by involving the various stakeholders that are involved. On the other hand, if it is true that climate change is a global issue,

Fig. 4.7 Art installation against irresponsible political approach against climate change, placed in Bonn, Germany, a few days before the 2017 United Nations Climate Change Conference (ph. L. Lenz, from Wikimedia Commons, licensed under CC0 1.0)

the management of its most evident effects, such as floods, coastal erosion and sea-level rise, is a responsibility of local governments who must interpret the objective and subjective factors of the phenomenon (Fig. 4.7).

References

Allen MR, Dube OP, Solecki W (2018) Framing and Context. In Vv. Aa. (eds) Global Warming of 1.5°C. An IPCC Special Report on the impacts of global warming of 1.5°C above pre-industrial levels and related global greenhouse gas emission pathways. Cambridge University Press, Cambridge, p. 49–92

Baiardi D, Morana C (2021) Climate change awareness: Empirical evidence for the European Union. Energy Economics 96

Belanche D, Casalo LV, Flavian C (2017) Understanding the cognitive, affective and evaluative components of social urban identity: determinants, measurement, and practical consequences. Journal of Environmental Psychology 50: 138–153

Benito G, Macklin MG, Zielhofer C, Jones AF, Machado MJ (2015) Holocene flooding and climate change in the Mediterranean. Catena 130: 13–33

Bergier T, Kowalewska A (eds) (2019) Addressing climate change in cities Catalogue of urban nature-based solutions. Ecologic Institute and Sendzimir Foundation, Berlin – Krakow

Bevasqua E, Maraun D, Vousdoukas MI, Voukouvalas E, Vrac M, Mentaschi L, Widmann M (2019) Higher probability of compound flooding from precipitation and storm surge in Europe under anthropogenic climate change. Science Advances 5(9)

Blöschl G et al. (2017) Changing climate shifts timing of European floods. Science 357(588): 588–590

Bollini L (2018) The Urban Landscape and Its Social Representation. A Cognitive Research Approach to Rethinking Historical Cultural Identities. In Amoruso G (ed) Putting Tradition into Practice: Heritage, Place and Design. INTBAU 2017. Lecture Notes in Civil Engineering. Springer, Cham, p. 834–842

Breil M, Catenacci M, Travisi C (2007) Impatti del cambiamento climatico sulle zone costiere. Quantificazione economica di impatti e misure di adattamento. CMCC Foundation, Lecce

Chauvin B (2014) La perception des risques: Apport de la psychologie à l'identification des déterminants du risque perçu. De Boeck Supérieur, Paris

Ciscar JC, Feyen L, Ibarreta D, Soria A (eds) (2018) Climate impacts in Europe. Publications Office of the European Union, Luxembourg

Doody P, Ferreira M, Lombardo S, Lucius I, Misdorp R, Niesing H, Salman A, Smallegange M, Cipriani LE, Lanza S, Pranzini E, Randazzo G (eds) (2004) Eurosion. Vivere con l'erosione costiera in Europa. Sedimenti e spazio per la sostenibilità. Publications Office of the European Union, Luxembourg

Evans E, Ashley R, Hall J, Penning-Roswell E, Sayers P, Thorne P, Watkinson A (2004) Foresight, Future flooding scientific summary: Volume II - Managing future risks. Office of Science and Technology, London

Field CB, Barros VR (eds) (2014) Climate Change 2014: Impacts, Adaptation, and Vulnerability. Part A: Global and Sectoral Aspects. Climate Change 2014. Working Group II Contribution to the 5th Assessment Report of the IPCC. Cambridge University Press, Cambridge

Gaglione F (2022) Città e climate change. La vulnerabilità delle aree urbane alle isole di calore. FedOA Press, Naples

Krien N, Michel-Guillou E (2014) Place des risques côtiers dans les représentations sociales du cadre de vie d'habitants de communes littorales. Les Cahiers Internationaux de Psychologie Sociale 101(1): 101–122

Lemée C, Guillard M, Fleury-Bahi G, Krien N, Chadenas C, Chauveau E, Desse M, Coquet M, Lamarre M, Navarro O (2019) What meaning do individuals give to coastal risks? Contribution of the social representation theory. Marine Policy 108

Lieske DJ, Wade T, Roness LA (2014) Climate change awareness and strategies for communicating the risk of coastal flooding: A Canadian Maritime case example. Estuarine, Coastal and Shelf Science 140: 83–94

Luo T, Zhang Z, Hong X, Wang Y, Zhang X (2023) Evaluating Spatial Identity Based on Climate Adaptation in Small Cities. International Journal of Environmental Research and Public Health 20(1)

Manigrasso M (2013) Città e Clima. Verso una nuova cultura del progetto. Sala Editori, Pescara

Masson-Delmotte V, Zhai P (2021) Climate Change 2021. The Physical Science Basis. Working Group I Contribution to the 6th Assessment Report of the IPCC. Cambridge University Press, Cambridge

Massoni ES, Barton DN, Rusch G, Gundersen V (2018) Bigger, more diverse and better? Mapping structural diversity and its recreational value in urban green spaces. Ecosystem Services 31: 502–516

Mendonça De Carvalho R, Szlafsztein CF (2019) Urban vegetation loss and ecosystem services: The influence on climate regulation and noise and air pollution. Environmental Pollution 245: 844–852

Michel-Guillou E, Meur-Ferec C (2017) Representations of coastal risk (erosion and marine flooding) among inhabitants of at-risk municipalities. Journal of Risk Research 20(6): 776–799

Moatti JP, Thiébault S (eds) (2016) The Mediterranean Region under Climate Change. A Scientific Update. IRD Editions, Marseille

Musco F, Fregolent L (eds) (2014) Pianificazione urbanistica e clima urbano. Manuale per la riduzione dei fenomeni di isola di calore urbano. Il Poligrafo, Padua
Musco F, Patassini D (2012) Mitigazione e Adattamento ai Cambiamenti Climatici: Valutazione di Efficacia di Piani e Politiche in USA, in Europa in Italia. In Pierabon A (ed) Nuovo manuale di Diritto e Gestione dell'Ambiente. Maggioli, Rimini, p. 809–830
Musco F, Van Staden M (eds) (2009) Local Governnments and Climate Change: Sustainable Energy Planning and Implementation in Small and Medium Sized Communities. Springer, Cham
Nakamura F (ed) (2022) Green Infrastructure and Climate Change Adaptation. Function, Implementation and Governance. Springer, Cham
Pecher A, Kofoed JP (eds) (2017) Handbook of Ocean Wave Energy. Springer, Cham
Pine JC (2014) Hazards Analysis. Reducing the Impact of Disasters. Taylor & Francis, Milton Park
Pörtner HO, Roberts DC (eds) (2022), Climate Change 2022: Impacts, Adaptation and Vulnerability. Working Group II Contribution to the 6th Assessment Report of the IPCC. Cambridge University Press, Cambridge
Reimann L, Vafeidis AT, Brown S, Hinkel J, Tol RSJ (2018), Mediterranean UNESCO World Heritage at risk from coastal flooding and erosion due to sea-level rise. Nature Communications 9
Rholi RV, Li C (2021) Coastal Flooding. In Rholi RV, Li C, Meteorology for Coastal Scientists. Springer, Cham, p. 309–315
Schleyer-Lindenmann A, Mudaliar R, Rishi P, Robert S (2022) Climate change and adaptation to coastal risks as perceived in two major coastal cities: An exploratory study in Marseille and Nice (France). Ocean and Coastal Management 225
Timmerman RJ (2021) Building Resilience Through Transboundary Water Resources Management. In Brears RC (ed) The Palgrave Handbook of Climate Resilient Societies. Springer, Cham, p. 97–115
Tubridy F, Scott M, Lennon M (2021) Managed retreat in response to flooding: lessons from the past for contemporary climate change adaptation. Planning Perspectives 36(6): 1249–1268
Van Lange PAM, Higgins ET, Kruglanski AW (eds) (2011) Handbook of Theories of Social Psychology: Geographical Perspectives. Sage, New York
Vecchio A, Anzidei M, Serpelloni E, Florindo F (2019) Natural Variability and Vertical Land Motion Contributions in the Mediterranean Sea-Level Records over the Last Two Centuries and Projections for 2100. Water 11(7)
Wu X, Zhang L, Zang S (2019) Examining seasonal effect of urban heat island in a coastal city. PLOS ONE 14(6)

Chapter 5
Governance of Coastal Public Space in the Euro-Mediterranean Context

Abstract The maritime public domain includes spaces and assets that serve navigation traffic, but this definition does not capture the wider interests of the coastal system. Europe, particularly the Euro-Mediterranean basin, has increasingly focused on social functions of the coast. The tourism industry, which exponentially developed in the last three decades of the twentieth century, has transformed coastal areas from places with a primarily logistical, commercial and military vocation to places for recreational uses. This shift is in line with the significant business potential generated by the city-sea interface but should also consider the community values inherent in coastal public spaces.

Keywords Maritime public domain · Coastal public space · Coastal governance · Bolkestein directive · Euro-Mediterranean basin · Maritime concessions

5.1 The Status of Maritime State Property from the Perspective of Social Uses

In order to understand the theoretical basis of the management for social use of the public spaces of the city-sea interface, it is necessary to define what is the meaning of maritime public domain, a controversial subject in the field of law since it can be assimilated to a *sui generis* sphere of the category of public goods. Generally speaking, maritime public domain corresponds to those spaces and assets connected with satisfying the needs of navigation and maritime traffic (Querci 1964). However, this definition appears limiting as it does not take into account the various interests that revolve around the coastal system in the broadest sense: in particular, it is appropriate to refer both to coastal areas usually affected to ordinary wave motion and to spaces that can be used for maritime activities related to navigation, fishing and bathing; therefore, two elements contribute to definine maritime state property, namely the naturalistic-geomorphological aspect of the shore and the functional aspect, connected to the public use of the sea (Ibba 2016).

I. Pistone, *Urban Coasts in Socio-ecological Transition*,
https://doi.org/10.1007/978-3-031-70783-4_5

In the Additional Protocol to the European Convention on Human Rights, the concept of *public property* is described from a communitarian perspective, indicating all the areas belonging to public state bodies, intended to satisfy certain public interests: this concept includes only non-disposable assets and state-owned assets (Council of Europe 2019). In particular, these latter consist of unmovable or registered movable property belonging to a territorial administration, like the state, regions or local authorities: depending on the circumstances, such property can be divided into necessary state property and potential state property. The former may be removed from private ownership altogether and pass into the hands of the territorial public administrations, for their intrinsic characteristics: these include military property, water property and, indeed, maritime property. On the other hand, the assets of the potential state property can belong to private owners, even though they can be included into the territorial public domain in case they are comprehended within the management sphere of a certain governmental body: this category can include, among others, various kinds of roads, railway lines, aqueducts, buildings of historical, archaeological and artistic value (Palma 1999).

On the basis of these concepts, it is possible to analyse the delicate issue of maritime governance in Europe, keeping in mind that the existing concession model presents numerous structural variants, not always favourable to an active fruition of the city-sea interface by the urban coastal society: indeed, urban shore is not even totally inclusive or prone to an easy accessibility and usability in many cases. First of all, the European model has different approaches in relation to the granting of contracts for public works, public services and public goods: for this reason, it is complex to organise the system of public concessions in an organic, disciplined and unitary way (Giannelli 2016). Actually, there is a great interest from the European Union in regulating other sectors not related to the allocation of public property licences: the effects of this line of thought are reflected in the uneven organisation in the management of the maritime state domain in the various European states (Caringella and Protto 2012). In fact, the first definition of concession system in EU law dates back to the *Public Works Contracts Directive* of 1993 (Directive 93/37/EC): in a certain sense, this directive assimilates the public concession to a project contract based on the realization of certain work for a price. Only in the new millennium, the *Directive 2004/18/EC* introduces the concept of public service concession, although it still gives priority to the coordination of public project contracts. The topic of public concessions gained progressive attention as a function of the gradual application of the community principle of equality: this should be the main aim towards which concession should tend, in order to preserve transparency in the attribution of licences.

Despite the regulatory criticalities at EU level, the social but also economic value of the maritime public domain clearly emerges, thus deserving an adequate attention in terms of urban and landscape planning. The complexity of this issue has been expressed by the European Directive 2006/123/EC (the so-called *Bolkestein directive*), which aims to enhance the free circulation of public services offered through maritime concessions. In this way, the directive tries to balance the dynamics underlying the concessions' system and the licence allocation process: the purpose is to standardise the coastal management system from an economic point of view and, consequently, in relation to the specific needs of coastal users. In particular, the Bolkestein directive (also known as the *Services Directive*) states that automatic renewals or extensions of maritime concessions for tourist-recreational activities are unconstitutional, even if there is a lack of adequate procedures for allocating licences among aspiring concession holders. This directive tries to examine all the potential situations in which the economic activity involves the use of limited or geographically circumscribed natural resources: in fact, it establishes that member states must organise a public call for possible candidates in order to allocate licenses, especially when the maritime concession is related to the management of scarce environmental assets, in order to guarantee impartiality and transparency in the procedure; moreover, it is necessary to clearly define the specific characteristics and the evaluation methodology of the call.

In any case, with Directive 2014/23/EU, the so-called *Concessions Directive*, the European Union seeks for the first time to regulate the issue of maritime concessions effectively and homogeneously. Concessions for public works, services and goods are a complex instrument: usually, the European Community has penalised the allocation of licences for public goods, probably because of an insufficient understanding of the relevant social implications (Garofoli 2019). Today, it seems that this aspect of the maritime public domain is still lacking a specific European regulation, also in light of the indications indirectly provided by the aforementioned Services Directive: as mentioned above, the main objective of Directive 2006/123/EC is in fact to guarantee the principle of freedom of establishment for service providers as well as to protect the free supply of services themselves; this clearly reflects the differences in the concession system in various European states, influencing certain socio-economic functions (Macchia 2018). However, from the point of view of the social use of the coast, Europe has shown a growing interest, especially in relation to the tourist industry that, starting from the last thirty years of the twentieth century, has enormously influenced the fruition of coastal spaces, changing the mainly logistic-commercial and military function into a more recreational and social use, coherently with the significant business volume that the city-sea interface can generate (Dezio 2016).

5.2 Governance in Europe: Managing the Mediterranean City-Sea Interface

It is now proposed an analysis of the governance systems within the main states of the Euro-Mediterranean basin, in order to understand the decision-making approaches in the geographical area covered by this work, concerning the management of the social uses of the city-sea interface. The studied countries (France, Spain, Greece, Croatia and Italy) possess a comparable environmental and social background: in particular, the Italian case will be described in an additional paragraph due to its unusual management complexity. From the geomorphologic point of view, these states present long coastal strips that are strongly urbanised and characterised by relevant socio-economic flows because of their landscape-environmental quality and strong tourist appeal; from the organisational and administrative point of view, their governance models for the maritime public domain appears to be articulated among multi-level competences and actors. With the analysis of the main institutional stakeholders involved in the maritime public land property, the distribution of responsibilities in attributing concessions and the main elements at the basis of maritime licences, the work tries to investigate how the Euro-Mediterranean area relates to the contemporary issue of satisfying the social needs of urban coastal society and the connection between public institutions and the environmental context of the city-sea interface.

- In 2006, the French management system repealed the *Code de domain de l'Etat*, dating back to the 1960s, in order to follow the indications of the *ordonnance 2006-460* establishing the *Code général de la propriété de personnes publiques* (CGPPP). According to the CGPPP, natural maritime public domain intuitively corresponds to the portion of coastline on which natural elements act and which has value for the satisfaction of social needs, considering state property unalienable and imprescriptible: in fact, although the concession of the aforementioned assets to third parties is contemplated, access to beaches and seashore in general must always be free of charge and openly granted to the community, for recreational, social, and economic purposes (such as fishing and aquaculture), except for reasons of safety, national defence or environmental protection.[1] In particular, with regard to beaches, the *Code de l'environnement* (*ordonnance 2000–914*) establishes that the state-owned area affected by such use must be left free of any structure or installation for at least 80% of the length of the reference shoreline. Moreover, permanent facilities on public coastal land are considered unacceptable: it is only allowed to install temporary equipment and structures, with no irreversible anchoring elements to the ground, coherently with the function of the state-owned property. This is linked to the French principle of restoring the original state of the places at the end of each concession, with a strong focus on environmental protection. Regarding the concession grant, the French administration follows the *Decree 2006–608 du 26 mai 2006 relatif aux concessions de plage*: this is a matter of state competence, administered by a specific prefect; the only exceptions are concessions located within the port domain, granted by the competent port authority. The duration of concessions extends to a maximum of twelve years: this short duration, together with the predominantly public management of the maritime state property, is unique within the Euro-Mediterranean governance framework. The concession drafting is subject to an open call procedure, in which the proposals are evaluated by the prefect, in case of both new contracts and renewals. Sub-concessions may be allowed, with regard to specific services that may also be monetised: anyway, their duration will not exceed

[1] The law of 31 December 1976 already established a buffer zone at least 3 mt wide for users to reach the sea: this element was later incorporated in the *Loi Littoral of* 1986.

the one of the main license. If the concession holder is a local authority, it is possible to follow a restricted procedure in which candidates are invited to submit proposals on the basis of a draft convention (on which the prefect's opinion is in any case required), in accordance with the *Code général des collectivités territoriales* (1996). It should be added that any occupation or use of the public domain involves the payment of a fee: the amount is established by the state authorities or local departments, depending on the situation, and is based on an assessment of the benefits that the license holder derives from the management of the public domain itself.

- In Spain, the Constitution (1978) includes the legal regime of the maritime public domain. Beaches are defined as free, so they are not subject to concessions: it is the state-owned area prior to the beaches themselves that can be subject to licences for authorising occupation. However, the Spanish legal system suffers from a neglectful urbanisation policy that in the last century has caused an irreversible alteration of the most valuable coastal areas, especially on the Mediterranean side. The gradual cementification of the coast was progressively caused by the lack of an efficient management system that would take into account, on the one hand, the protection of the environmental heritage, which citizens should freely enjoy, and, on the other hand, the demands of economic and recreational development: in fact, Spanish seasonal tourism generates a flow of almost 40 million visitors annually, in comparison with 7 million inhabitants living in a coastal strip only 4.5 km wide, according to the data provided by the INE (Instituto Nacional de Estadística). The reference law for the governance of the coastal interface is the *Ley de Costas* (Law 22/1988). When this law was approved, controversies emerged about private buildings on the coast that had been transferred to the public domain: for this reason, the Spanish state granted the interested parties a 30-year public concession in return for an annual fee, without renouncing to the public ownership of the areas. The *Ley de Costas* provides for two types of licences for the use of the maritime public land property, namely authorisations and concessions. The authorisations refer to facilities and activities regulated by the general rules of the legal text, they have a four-year duration and their request can be subject to public promotion, according to the specific normative: in this category, it is possible to include profit-making activities that can take place on the beaches, also through removable elements, such as renting sunbeds and beach umbrellas; the space for these activities must not exceed half of the reference coastal area. For concessions linked to the use of coastal public domain, the allocation process generally (but not exclusively) follows a competitive procedure, although the call and granting parameters are decided by the reference administration.[2] The duration of the concession-based relationships depends on the specific use of the maritime domain and it has to be fixed on a case-by-case basis, although it is set

[2] The allocation of concessions does not necessarily follow the institution of a public call: it should be noted that if a user has previously expressed its interest in the management of a part or element of the coastal state domain and is not awarded with the relative licence, he may still request the cost coverage for his proposal, if this is provided for by the call regulation; moreover, the public call may be considered cancelled if none of the participants has adequate qualifications.

at a maximum period of seventy-five years, which can be extended in exceptional circumstances for a further seventy-five years, according to the provisions of *Ley 2/2013 de protección y uso sostenible del litoral* (2013). This law also responds to the imminent expiry of the thirty-year concessions of 1988, effectively extending them. With regard to the transferability of maritime concessions, the Spanish regulation allows the transfer of licences between living stakeholders and admits *mortis causa*: in case of demise of the concession holder, his heirs may take over the management of the portion of maritime public domain that is regulated by the reference concession: for this to happen, they have to submit a special application form within four years from the death of the original concession holder, otherwise the license itself will expire.

- The situation in Croatia is characterised by considerable attention to the preservation of natural resources and sustainable development, which is certainly linked to less economic and speculative transformation along the national coastline. Maritime state property is mentioned in Article 52 of the Constitution (1990): the sea, beaches and islands shall be subject to special protection, because of their great value for the Croatian Republic. The sectoral legislation codifies the appropriate exploitation of maritime public domain by concession holders, as well as the specific functional restrictions. The regulation of the uses of maritime state property refers to the *Maritime Domain and Seaports Act No. 11/2002*. This act classifies all natural elements that fall within the notion of maritime public domain; it also establishes the different authorities regarding the coastal protection, which is national responsibility, and the public domain management, which is administered by both regional and state bodies. The Maritime Domain and Seaports Act precisely states the dynamics for granting concessions, clearly expressing a dual category. Concessions aimed at the economic exploitation of the maritime public domain must follow a public tender procedure, as they allow the concession holder to enjoy a potential financial profit through a public asset. On the contrary, concessions for special uses are granted upon the request of the interested parties: special uses are those socio-cultural and sports activities that originate from a private initiative but give benefits to the whole community, which is why in this case the license allocation is untied from the competition procedures required by the European Union; in any case, if there are multiple requests, the tender procedure for the granting of the concession follows the decision of a group of experts, selected by the central government, the county concerned or the municipal administration, depending on the relevance of the intervention. Regarding the potential works on the maritime public property under concession, the general principle is to prefer light and detachable structures, so that they can be easily removed upon termination of the concession-based relationship, preserving the environmental status of the shore as much as possible; however, if it is required to install permanent facilities, they become part of the maritime public domain on which they stand and their ownership is transferred to the public body. The duration of licences varies and ranges between five and ninety-nine years, although there are precise indications to establish the right time period of the various concessions, on a case-by-case basis.[3]

- Greece refers to *Law 2971/2001* for the regulation of the maritime public domain. This law promotes ideals of transparency and impartiality for authorising recreational and tourist activities along the beaches, allowing the possibility to lease certain areas of the public land property in order to exploit the shore for public utility purposes. On the other hand, as far as accommodation facilities next to beaches are concerned, the management system provides for exceptions to allow hotels to obtain annual authorisations in order to carry out their activities. It also forbids permanent constructions within 100 m from the shoreline, as well as related purchase deeds and any other changes to the urban territory, unless a special administrative measure has been adopted to delimit the water edge or if there are ongoing environmental recovery interventions on the beach or the coast. It should be noted, however, that Ministerial Decree 5159/586 of 2014 has introduced direct awarding of beach concessions by municipal authorities: in this way, it has been possible to approve 80 project proposals related to tourist and recreational uses of beaches protected by the NATURA 2000, in accordance with the Prioritised Action Framework For Natura 2000 For The EU Multiannual Financing (European Commission 2014). At the same time, the law states that accommodation facilities have to apply for special licences in order to install temporary equipment for bathing activities, such as floating wooden docks or beach chairs for customers. Moreover, the duration of licences for tourist ports has been extended. It is thus clear that the Greek maritime domain governance is sometimes heedless of the public character of coastlines and beaches. However, a progressive attention is emerging towards the protection of littoral zones, also limiting the perimeter of areas to be managed with maritime concessions, in order to avoid overcrowding phenomena within the bathing establishments.

5.3 Coastal Maritime Concessions in Italy: A Peculiar Case in the Euro-Mediterranean Basin

The Italian coastline stretches for about 8300 linear kilometres, with 3346 km of beaches. In this area, there are about 61,500 maritime concessions, even if less than half can be linked to social and recreational uses: of this important fraction, about

---

[3] According to Section 20 of the Maritime Domain and Seaports Act, concessions linked with the economic exploitation of the public domain and the realisation of relevant buildings at a regional level may last up to twenty years: this period may be extended to a maximum of thirty years in relation to any new funding or due to force majeure. The license holder has to request the concession extension and needs the approval of the central government. Only if the proposal on the concession area has national resonance, the licence may be valid for up to fifty or sixty years.

12,000 concessions are related to bathing establishments, while almost 1900 concessions concern campsites, sports areas or tourist facilities[4]; the origin of the licences is varied but mainly connected to Regions, Municipalities and Port Authorities. These data anticipate the complexity of the concession topic in the Italian maritime domain within the Euro-Mediterranean panorama, with regard to the allocation of competences, the time span of licences and the free use of public spaces along the city-sea interface. The concept of maritime state property is defined in the Navigation Code (*Codice della Navigazione*, R.D. 327/1942), which establishes what the public domain actually is[5]: this categorisation also includes territorial sea, in relation to public uses and the satisfaction of collective needs, and port areas, which constitute another form of state property, defined as artificial. Moreover, it is introduced the possibility of drafting concession contracts for the occupation and use of the aforementioned state land property.

In relation to social uses of the city-sea interface intended to meet the needs of urban coastal society, it is relevant to note that the first codification of maritime concessions with explicitly tourist-recreational purposes dates back to Law 494/1993 for the determination of fees for public licenses: this law indicates the categories of compatible public services, from the management of bathing establishments, accommodation facilities and recreational-sports activities to catering establishments and boat renting. These regulatory indications were then merged into Law 296/2006 (*Legge Finanziaria del 2007*), which defines when licenses could be granted for tourist-recreational uses, with a distinction from concessions for docking points and recreational boating facilities, in relation to fees to pay.

Since the 1970s, however, the need for efficient coordination in the dynamic management of maritime domain has emerged from an economic but also a landscape and environmental point of view. The Presidential Decree 616/1977 areas started a process of gradual decentralisation of certain administrative functions for the grant of concessions, giving more responsibilities to Regions. In particular, the central government holds control over the property aspects, the functions related to national security and maritime navigation and the energy supply. On the other hand, the Regions have the task to manage the tourist and recreational areas of the public domain along the maritime coastline: however, the implementation of this delegation has been considerably delayed and only in 1990s the Prime Ministerial Decree 21/12/1995 for the identification of maritime state-owned areas and Law 59/1997 (*Legge Bassanini*) were drafted to simplify the administrative procedure for granting licences. Under these premises, the Regions then autonomously started to regulate the features of the concession-based relationships: a further decentralisation has become a common practice in order to give municipal authorities the power to

[4] Data provided by the Ministry of Infrastructure and Transports through the Maritime Domain Information System (SID) (https://www.sid.mit.gov.it).

[5] In Article 28 of the *Codice della Navigazione*, the following elements are included in the maritime domain: '*the shore, the beach, the ports, the bays; the lagoons, the mouths of rivers flowing into the sea, the basins of saline or brackish water that freely communicate with the sea at least during part of the year; the channels that can be used for public maritime uses*'. It is also stated that structures insisting on state-owned areas effectively become part of them.

assign the actual concession licenses and to eventually modify them, always in a recreational context. In addition to this, municipalities are entrusted with the collection of fees, calculated on the basis of surface extension, the ratio between covered and uncovered areas, and the distance from the sea, in accordance with the aforementioned *Legge Finanziaria del 2007*. Another parameter to consider is the eventual realisation of facilities insisting on the maritime domain, in order to ensure urban, architectural and landscape-environmental conformity between the interventions on the city-sea interface and the existing plans: indeed, checking this compatibility is usually a municipal responsibility.

From the planning point of view, the Regions have progressively approved several plans for the use of state-owned areas in order to ensure their tourist and social exploitation. They usually provide guidelines for the drafting of Beach Plans on a local scale, also regulating detailed parameters, such as the dimensional and spatial features of the buildings on the coast. Municipal planning of the shoreline tends to intertwine with the town plans: in fact, the latter manage areas that are not strictly connected to the perimeter of maritime concessions but nevertheless relevant to the social use of the urban amphibious. Local planning usually embodies the identification of functional zones into state-owned coastal areas, in order to guarantee spaces for free transit along the shore, areas for resting in the shade and for beach services, as well as a clear area to enjoy the coast without paying any charges (Carlin 2019). It should be noted, however, that the integration of state concessions and coastal planning can often cause bureaucratic slowing, as specific permissions may be subject to approval delays pending the implementation of the aforementioned Beach Plans.

On the other hand, with regard to tender procedures for licenses assignment, there is a lack of regulations in the Italian case, even though public domain management is a very topical subject because of the considerable profit opportunities it may bring to entrepreneurs. In spite of the fact that the European Union has dedicated significant efforts to regulate the drafting of concession-based relationships since the early 2000s, in Italy any specific regulation has been approved yet in order to define modalities, times and terms of public calls and the evaluation of tenders. At the national level, the criteria of the Navigation Code are still in force, even if they are extremely generic about the content of calls for tenders and the consultation process for approval: this decades-long gap has prompted the regions to take decisions on their own, devoting greater or lesser attention to the issues of public participation, adequate promotion and identification of evaluation criteria, depending on the sensitivity of the lawmakers. Such uncertainty translates into not always equal treatment among participants in public calls organised by the municipal administrations: the reason lies in the high discretionary power attributed to the judging commissions and in the difficulty for potential concession holders in precisely calibrate their proposals (Giannaccari 2021a).

Another critical issue concerning concessions along the Italian coastal interface is the duration of licences: the time parameter initially adopted by the Navigation Code was vague, not providing specific indications on the maximum time frame to be considered. The Law 88/2001 (*Legge 16 marzo 2001, n. 88. Nuove disposizioni in materia di investimenti nelle imprese maritime*) set the duration of concessions at six years, but then confirmed automatic renewal upon expiry of the licence, confirming the right of insistence already defined by the Navigation Code. The right of insistence (which will only be abrogated by Decree-Law 194/2009) is a thorny issue because it legitimises administrations to reconfirm the concessions already granted during the process of license renewal, nullifying fair competition. The motivation for this choice lies in the stability of concessions that Italian legislation has pursued in the field of coastal management, so that entrepreneurs could adequately equip the shores in the long term, increasing private earnings and strengthening the attractiveness of the coastal interface for the tourism industry and for potential new investments (Lami et al. 2010).

Predictably, this criterion was opposed by the European Commission, which initiated the *Infringement Procedure 2008/4908* against Italy, closed only in 2012, since the right of insistence is at odds with the principles of the Bolkestein directive, according to which it is not allowed to grant automatic renewals, also due to possible corruption of public administrations and illegal agreement with concession holders. During this time, the *Legge Finanziaria del 2007* set a minimum duration of six years and a maximum duration of twenty years for concession-based relationships, which can only be reached in the case of major works or investments, in accordance with regional plans for the use of public domain and municipal urban planning instruments. In any case, the Italian State faced the issue again, first with the aforementioned *Decree-Law 194/2009*, granting an indistinct extension to all maritime concessions with recreational tourist purposes until December 31st, 2015, and subsequently with *Decree-Law 179/2012*, granting a new extension until December 31st, 2020 and expanding the categories of services to be managed by the licence (including docking points for recreational boating). *Law 145/2018*, which became effective in January 2019 as the *Legge di Bilancio*, finally declared that the aforementioned maritime concessions would remain active until 2033 (Giannaccari 2021a, b). These events were debated in the Plenary Meeting of the Council of State (*Judgments of the Plenary Assembly no. 7 and 18 of 2021*), which put a brake on the extensions of licences as incompatible with Directive 2006/123/EC, limiting the duration of the reference concessions to December 31st, 2023, although this deadline has been further postponed by one year by Law 18/2024, following the *Decreto Milleproroghe*.

5.4 Reasoning About the Mediterranean Coastal Management

In spite of these reflections, however, it must be noted that the issue of governing the maritime public domain is far from a Community resolution. There is still an unresolved discontent regarding the application of the Bolkestein directive from a recreational-bathing point of view, the alleged governance instability and the presence of normative constraints in the license transfer from one concession holder to another; these are economic but also social issues, since the relevance of the coasts, especially in urban areas, is difficult to quantify. There is need for at least greater clarity in the topic of management of coastal state property (di Plinio 2020). This short descriptive framework is therefore intended to highlight the governance complexity of the city-sea interface, since it is necessarily a multi-stakeholder context but it is also inevitably monopolised by interests that often diverge from the free accessibility and use of public space.

However, a study of the different Euro-Mediterranean management frameworks highlights both positive and negative aspects. On the one hand, there is a progressive interest on the part of the competent authorities in the integrated management of the coast, associated with the protection of environmental qualities: in this context, France undoubtedly stands out, efficiently transposing Community directives and making them operational, also by limiting permanent installations on coastal land (imposing, for instance, the removal of bathing facilities at the end of the summer season). Moreover, it has to be noted the gradual introduction of various public tender systems to liberalise the development of economic capital, in order to ensure an adequate fruition of coastal services, in accordance with European indications: however, competitive behaviours may also lead to an excessive monetisation of the maritime public domain and its assets, because enhancing functions specifically dedicate to public sociality could bring economic disadvantages to the concession holders; anyway, if the central authorities keep control of the cultural and landscape quality of the urban amphibious, the management and allocation of licences for coastal exploitation is often delegated to regions and municipalities, entrusting private users with the implementation of projects and facilities (Benetazzo and Gobbato 2017).

The enjoyment of urban coasts implies the active involvement of all the users of coastal society, suggesting a reorganisation of the granting procedure of concessions, avoiding to let few private individuals in charge of coastal management and directing the process towards integrated governance systems, with the participation of communities, local authorities and associations, or even returning the governance of the coastal territory entirely into the hands of the public institutions, in order to grant the community aspect of the coastal state assets (Lucarelli et al. 2019). If it is true, however, that the services that the coastal landscape offers are linked to recreational and social activities for the public, it should also be considered that they require the realisation of various facilities within the maritime domain and the costs of these interventions are generally sustained by the concession holder. In this sense, the analysis of the Italian case aims to show the existing contrasts between the beach service suppliers and the central government, which has been sanctioned for the continuous delays in implementing the Bolkestein directive. However, the issue appears to be very complex because it is related to the difficult landscape and cultural enhancement of the coastal interface, from an economic and tourist point of view, but also to the actual obstacles in the management of state-owned littoral space for small and medium-sized entrepreneurs often involved, in a national, regional and local context where planning and administrative efficiency is generally lacking. A more articulated dialogue with European institutions would therefore be useful in order to find an adequate balance (Gola 2021); moreover, urban coastal society should keep a close interaction with the city-sea interface, especially because of the very strong socio-cultural value linked to urban coasts.

Indeed, in almost all of the examined cases, the service provider has to deal with the request for the planning permission and landscape authorisation needed to install the necessary equipment for ensuring coastal use by the littoral community: since he only has surface right, the built assets can be transferred to the maritime public domain at the end of the concession. This is a matter of concern: on one hand, it is hoped that the license holders decide for light and detachable elements instead of permanent ones, guaranteeing the restoration of the original condition of the coastal areas by the end of the licence, with obvious advantages for the protection of the coastal landscape value; on the other hand, it should be noted that it is not always feasible to choose temporary installations due to the land composition, atmospheric agents and other causes that impose more resistant and definitive design solutions for safety reasons (Calabrò 2021). It is therefore necessary to think carefully about achieving an integration between management tools for the maritime public domain and the urban plans existing along the city-sea interface: it is compulsory to resolve the conflicts between public property and private property, guaranteeing a careful approach to preserve the environmental features of the shore and to ensure a fair and transparent fruition by the urban coastal society, while managing these characteristics flexibly and coherently with the changing legal and socio-ecological conditions of the urban amphibious (Provenzano 2011).

References

Benetazzo C, Gobbato S (2017) Concessioni balneari in Italia e Direttiva 2006–123-EC, nel contesto europeo. Directorate-General for Internal Policies, European Parliament

Calabrò M (2021) Concessioni demaniali marittime ad uso turistico-ricreativo e acquisizione al patrimonio dello stato delle opere non amovibili: una riforma necessaria. Diritto e Società 3: 441–472

Caringella F, Protto M (2012) Manuale di diritto processuale amministrativo. Dike Giuridica Editrice, Rome

Carlin M (ed) (2019) Concessioni demaniali marittime e lacuali. Problemi e casi pratici. Key Editore, Milan

Council of Europe (2019) European Convention on Human Rights

Dezio G (2016) Le concessioni demaniali alla luce delle recenti prospettive di riforma. Il Diritto Amministrativo 17

di Plinio G (2020) Il Mostro di Bolkestein in spiaggia. La "terribile" Direttiva e le concessioni balneari, tra gli eccessi del Judicial Italian Style e la crisi del federalizing process. Federalismi 2020

European Commission (2014) Prioritised Action Framework For Natura 2000 For The EU Multiannual Financing

Garofoli R (2019) Manuale di Diritto amministrativo. Neldiritto Editore, Bari

Giannaccari A (2021a) À la guerre comme à la guerre. Concessioni demaniali marittime, Adunanza plenaria e procedure selettive (al 2023?). Mercato Concorrenza Regole 3: 581–591

Giannaccari A (2021b) Stessa spiaggia, stesso mare. Di concessioni demaniali marittime e (assenza di) concorrenza. Mercato Concorrenza Regole 2: 307–341

Giannelli A (2016) Beni sfruttabili o consumabili: demanio marittimo e porti. Federalismi 22
Gola M (2021) Il Consiglio di Stato, l'Europa e le 'concessioni balneari': si chiude una – annosa – vicenda o resta ancora aperta? Diritto e Società 3: 401–418
Ibba FG (2016) Le concessioni demaniali marittime ad uso turistico-ricreativo alla luce dei principi dell'Unione Europea, Ph.D. thesis, University of Sassari, Sassari
Lami G, Nebbia Colomba CA, Villamena S (2010) Le concessioni demaniali marittime. Tra passato, presente e futuro. Exeo Edizioni, Padua
Lucarelli A, De Maria B, Girardi MC (eds) (2019) Governo e gestione delle concessioni demaniali marittime. Edizioni Scientifiche Italiane, Naples
Macchia M (2018) La regolamentazione degli affidamenti da parte dei concessionari. Rivista Italiana di Diritto Pubblico Comunitario 26(1)
Palma G (1999) Il regime giuridico della proprietà pubblica. UTET, Turin
Provenzano S (2011) Il PUDM come occasione di progetto del waterfront urbano: il caso di Sant'Agata di Militello (Me).Portus plus 1: 1–10
Querci EA (1964) Il demanio. In Enciclopedia del Diritto 12. Giuffrè editore, Milan

Chapter 6
Research Question: Socio-Ecological Use of the City-Sea Interface in the Euro-Mediterranean Context

Abstract The objective of the work is to analyse the regeneration process of city-sea interfaces by investigating the neuralgic elements and communities inhabiting them. The main research question addresses how to support urban planning in enhancing existing or potential coastal public spaces for socio-recreational and environmental benefits. Thus, it is proposed a methodological framework to analyse legal, geomorphological, and psycho-social aspects of city-sea interfaces. The major target are large coastal cities, in particular along the Euro-Mediterranean sea, due to the presence of complex challenges, stakes and relevant socio-demographic variations that urban planning must consider for improving the overall littoral quality.

Keywords Methodological framework · Coastal public space · Mediterranean sea · Large coastal cities · Coastal planning and management

6.1 How to Support Planning and Management of Coastal Public Space from a Socio-recreational and Environmental Perspective?

The reflections that have defined the field of research indicate the need to investigate the neuralgic elements of the urban coasts by looking at the communities that inhabit them, thus rehabilitating the urban scars fragmenting the city-sea interface, restoring the balance between identity value and functional value of these peculiar areas of the contemporary city. In particular, the interest focuses on those spaces that most of all allow the user of the urban coastal society to enjoy the coast and the sea, in a social and recreational way: the public areas along the urban shorelines can in fact grant well-being and the possibility to socialise, to relax, in direct contact with the most relevant environmental resource of coastal cities: the sea. The relationship with water in the city is essential for urban coastal society; anyway, the fruition of this natural resource through urban beaches, public parks, blue-green areas and coastal open spaces in general is not always equal and immediate, also because of a certain management complexity; in the same way, recreational functions are present in the

I. Pistone, *Urban Coasts in Socio-ecological Transition*,
https://doi.org/10.1007/978-3-031-70783-4_6

sea too, for instance sport or non-intensive fishing, diving, cultural activities related to underwater heritage, bathing. In addition to responding to the growing social demand for public spaces for social purposes, even if these areas have a relatively limited surface, they still possess the potential to react adaptively to the effects of climate change and environmental impacts, through flexible and multifunctional design: an efficient integrated organisation is therefore required for these zones of the city-sea interface, which, as mentioned above, is comparable to an amphibious entity in terms of both uses and planning tools.

In this sense, it is appropriate to investigate the following research question: how to support urban planning and governance processes in intervening in existing or potential coastal and maritime public spaces in order to mend the fragmented structure of the city-sea interface, from a socio-recreational and environmental perspective, allowing users equal access to the urban sea and thus increasing the quality of life in large coastal cities?

The research question therefore emphasises the need to return quality public spaces to citizens, considering the city coastline as a connecting platform for urban coastal society. The community can indeed actively participate in the improvement of the city-sea interface, where projects and interventions can be drafted in a flexible but organised manner, channelling planning innovation into a democratic interactive perspective. Inclusion and social cohesion motivate users to participate in the improvement of the interface, in favour of a renewed sense of place that is crucial to overcome social barriers. Restoring urban beaches, the realisation of new natural parks along the coasts (but also the implementation of sustainable socio-recreational uses of urban waters), together with the participative relationship among the users of the urban coastal society, are part of the research perspective: these areas, even if small in size, can in fact become the socio-environmental engine of the city-sea interface, on condition that they are proficiently planned and connected among themselves, through social and urban design innovation. Hence, a functional reorganisation of coastal public spaces sounds plausible, in accordance with the current socio-recreational demand that is less oriented towards large public spaces dedicated to tourists and visitors and instead emphasises more and more the importance of proximity public spaces that allow citizens to enjoy their city on a smaller scale.

Consequently, the need for an integration of coastal design with water plans, down to the local and micro-local scale, is investigated. The principles of Integrated Coastal Zone Management (2000) suggest a broad holistic perspective that recognises the interactions among flows and elements in the littoral areas in relation to hydrological, geomorphological, socio-economic, institutional and cultural characteristics, considering participatory planning as a way to put together local and natural specificities. Maritime Spatial Planning (2014) proposes adaptive management and the implementation of plans for water and coastal use, taking into account land-sea interactions and environmental, social and safety aspects: however, only a limited number of European countries have implemented such tools until today. The 2016 New Urban Agenda promotes the ecological and social functions of public spaces to ensure a progressive improvement of urban quality, focusing on ecological sustainability and design. The functional recomposition of the fragments of the city-sea interface can thus amplify the coastal value because it may help understand spatial and ideological conflicts through innovative approaches that get to the root of the problem, trying to solve the issues between the urban fabric, the coastal society and its territory. In particular, the mosaic of natural and man-made spaces of urban coastal areas can also be defined as an articulated hydrographic network: in this sense, the green–blue infrastructure can ecologically restore the territorial environment, guaranteeing the fruition of ecosystem services and limiting hydraulic risk and coastal erosion, for whose prevention the approach was originally conceived.

Besides the theoretical contribution, the volume also aims to outline a methodological framework that can support planners and inspire policy makers regarding the legal, geomorphological and psycho-social characteristics of the city-sea interface for an efficient socio-environmental urban planning and management, in response to the criticalities highlighted for the urban amphibious. The analysis of the geographical, spatial and social context can in fact allow efficient solutions based on adaptive processes, identifying priorities and impacts of coastal interventions. As a result, it is possible to adopt a multidimensional and perceptive approach that studies urban coastlines according to the shared understanding of littoral public space that different social groups manifest, studying which coastal uses are most appropriate, the importance that citizens attribute to these areas and the influence that the environmental system has on the usability of open coastal spaces, in order to enhance physical and visual accessibility to maritime amenities.

6.2 The Relevance of the Euro-Mediterranean Case Study

Today, rethinking how to plan and manage the coastal public space is a common topic in different urban realities: however, it should be noted that each urban coastal society is unique and it is not possible to approach each city-sea interface in the same way; instead, it is necessary to investigate a diagnostic approach that studies the variables involved in urban coasts in order to discover their common features, from a planning and administrative perspective. It makes sense to analyse what are the main aspects that characterise the spaces of direct and indirect interaction with coastal water, in relation to the progressive coastal urbanisation and the impacts on urban blue spaces.

In particular, the book will focus on the Euro-Mediterranean basin. As described in the introduction of this text, the Mediterranean area represents an absolutely unique functional zone in the world. From an environmental point of view, the coasts are under great pressures, due to strong urbanisation and growing climatic risks that threaten both ecological ecosystems and the safety of urban coastal society, effectively limiting the usability of the city-sea interface; furthermore, Mediterranean cities share a strong bond with the natural and built landscape, as a result of an articulated cultural background that somehow unites the settlements along the coasts of different states. It is also an area where, over the centuries and particularly in the last fifty years, several large urban centres with a high population density have started

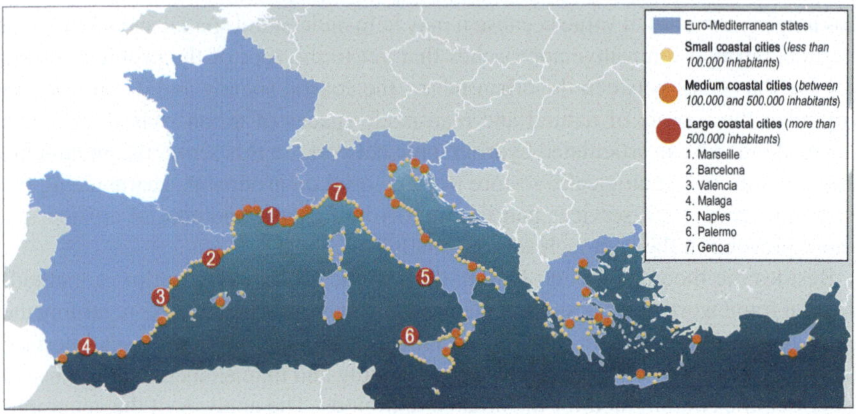

Fig. 6.1 Location of the small, medium and large urban settlements on a demographic basis within the Euro-Mediterranean basin (elaboration of the author, based on Eurostat data)

to develop more and more: in these cities, the intensity of the socio-environmental relationships along the city-sea interface is particularly high. However, these cities need to adopt different approaches to grant equal fruition of the public space along the littoral interface, due to the diversified planning and legal frameworks of the reference countries: this demonstrates the importance of understanding how to flexibly adapt the methodological approach to the city-sea interface according to the geographical and social context.

Specifically, large cities will be the main objective of this work, since there the criticalities listed so far are amplified, due to the numerous stakes involved, the complex functional articulation of their city-sea interfaces and the demographic variety of the local urban coastal societies. In the Mediterranean context, the urban coasts of large cities in the European countries generally share a high population density, while the area usually appears connected to large port infrastructures. Despite this, it is possible to find places of social transformation along the urban amphibious, with some attention for climate change too. In order to study the most varied conditions of littoral urban development in relation to the social demands of urban coastal society, the range of investigation will focus on cities with more than half a million inhabitants: specifically, Marseille and Naples will be examined as in-depth case studies of the research work, coherently with the changing legal and socio-ecological conditions of the urban amphibious (Fig. 6.1).

Part II
The Methodological Framework: Programmatic and Design Approach to the City-Sea Interface

Chapter 7
Maritime Spatial Planning: Normative Analysis of the Euro-Mediterranean Coast

Abstract The chapter investigates the main EU directives regarding coastal areas, aiming to assess the European community's consideration of socio-environmental development along urban coasts, particularly in the Mediterranean. Thus, the Maritime Spatial Planning (MSP) framework is pertinent, offering a project-oriented approach to coastal management. MSP involves analysing and efficiently allocating maritime functions and uses in order to minimise conflicts among coastal stakeholders and maximise benefits from human activities, all while safeguarding marine ecosystems. It is connected to land-sea interactions, amplifying the complexity of urban littoral planning and intertwining with the study of public spaces along the coastal interface.

Keywords Maritime spatial planning · Integrated coastal zone management · Marine strategy framework · Land-sea interactions · Local scale

7.1 EU Directives for the City-Sea Interface

For the normative and legal aspect of this work, it is necessary to clarify that there is not a unique way of interpreting maritime space, since its delimitations derive from different directives and uses, depending on location, governance models and local coastal community. In fact, the boundaries of maritime and coastal areas tend to gradually dissolve: despite this, they generate considerable interest in the study of EU legislative aspects, from a social as well as an economic-environmental perspective, on the basis of the benefits they produce. It is therefore correct to say that urban coastal areas are progressively losing their role as a border, as a demarcation line between different systems (the city and the water) to fall into the category of interface, as explained above. For these reasons, an examination of the most relevant recent directives on the subject of coastal waters at a supranational level is proposed, in order to establish the extent to which the socio-environmental development of urban coasts is considered by the European community and, in the specific case

I. Pistone, *Urban Coasts in Socio-ecological Transition*,
https://doi.org/10.1007/978-3-031-70783-4_7

of the present volume, in the Mediterranean context. This study is essential as the city-sea interface clearly has a dual nature, terrestrial but also maritime.

Firstly, the *Recommendation of the European Parliament on Integrated Coastal Zone Management* (ICZM) (2002) recognises the great environmental, socio-economic and cultural importance of the European coasts, while highlighting, already twenty years ago, the threats posed by climate change and the progressive increase of demographic pressure on urban coasts; it suggests a broad holistic perspective in order to recognises the interconnection of hydrological, geomorphological, socio-economic, institutional and cultural characteristics in coastal zones: within these elements, participatory planning could be an instrument of synthesis between local and natural specificities (Morales 2022). Indeed, the ICZM Recommendation identifies that the increasing pressure of human activities, together with the impacts of climate change, can seriously threaten the functional integrity of coastal ecosystems, requiring a multidisciplinary and multi-actor approach to understand the main drivers of change, but also the pressures on European coastal areas: hence, there is a need for an effective integration of ecological, social and planning data and expertise in policies and management in order to pursue a balanced and sustainable development of coastal spaces (Moksness et al. 2013). In any case, from a public policy perspective, there is not a single and shared definition of coastal zone, as it is an area defined according to the peculiar environmental and administrative characters of the different reference countries: the purpose of Integrated Coastal Zone Management plans is therefore to administer activities and uses that directly or indirectly cover the space between land and sea, as coastal interactions are linked to environmental or socio-economic systems that affect both the terrestrial and the maritime territory (Ahlhorn 2018). In particular, the document opens up for an ecosystem approach that can protect the coastal zone from environmental hazards, working towards the strengthening of an efficient socio-cultural system for littoral society; it is interesting that the directive indicates the need to provide adequate accessibility to the public coastal space, according to aesthetic and recreational characteristics, while recognising the interdependence but also the discrepancies between the natural coastal system and human activities.

The *Marine Strategy Framework Directive 2008/56/EC of the European Parliament and of the Council*, better known as the *Marine Strategy Framework Directive* (MSFD) (2008), introduces an important perspective on the environmental quality of coastlines, focusing on the relevance of the ecological assessment of the water conditions in relation to the impacts generated by human activities along the coast, reflecting on the implications for marine biodiversity, habitats and ecosystems. In fact, the directive aims to enhance a better understanding of the anthropogenic pressures on the sea, in order to pursue a valid and sustainable use, also encouraging collaboration between the various European coastal states and between the various smaller-scale maritime regions, working towards shared strategies to face anthropogenic changes in the coastal territory. The MSFD explicitly requires member states to take social and economic aspects into account in the preparation and implementation of their marine strategies, through the evaluation of the sea status and socio-economic analyses of its use, considering the eventual degradation of local environments. Indeed, ecosystem services provide uncountable benefits to urban coastal society, but their protection comes at a price: consequently, the directive calls for estimating the opportunity cost of measures to preserve maritime-coastal ecosystems through a careful cost–benefit analysis of the different potential policies, in order to select the most convenient choice, comparing environmental defence actions with other more damaging investment opportunities (Bertram and Rehdanz 2013). This led, in May 2020, to the adoption of the *EU Biodiversity Strategy for* 2030: its objective is to increase protection of the marine system in order to preserve an acceptable environmental status through the management of protected areas and the spatial planning for sustainable social and economic use of the coastline, paying attention to the sources of pollution related to tourism and recreational activities.

In 2014, *Directive 2014/89/EU* of the European Parliament establishes a framework for *Maritime Spatial Planning* (MSP): it is particularly relevant because it provides a specifically project-based approach for the issue of coastal management. It is a complex and strategic process for analysing and efficiently allocating the uses of maritime space in order to minimise contrasts between different actors along the coast and maximise the benefits of human activities, while ensuring the resilience of marine ecosystems. Indeed, the directive emphasises the rapidly increasing demand for maritime space for diverse purposes related to renewable energy production or fossil fuel and natural gas exploitation, maritime transport, fisheries and aquaculture, tourism and archaeological-cultural heritage: these features require an integrated planning and management approach (Fig. 7.1).

Maritime Spatial Planning, taking up the indications of the previous EU directives, aims at proposing an interdisciplinary policy tool that offers the possibility to deal with sea-related issues in an ecosystem-based way, promoting sustainability as a development engine for the maritime and coastal economy and as a guiding principle for a responsible fruition of sea and coastal resources in general. The instrument is addressed to a very broad range of users and covers different functional sectors, according to the cumulative impacts they may generate, providing spatial and time measures for an efficient fruition of marine areas or resources (Iglesias-Campos et al. 2021). It also mention the effects that natural disasters and climate change can have on coastal dynamics, influencing littoral human activities: attention is paid to the effect that phenomena such as the shore erosion and coastal advancement due to sedimentation have on the positive growth of the wider coastal-maritime ecosystem, in terms of damage on the ecological status, loss of biodiversity and degradation of ecosystem services; these elements must be a cornerstone for the elaboration of maritime spatial management plans, as their integration within the planning process can bring considerable benefits in relation to recreational and tourism activities, adaptation to climate change and control of coastal zone dynamics and disasters (Douvere and Ehler 2011).

Specifically, it proposes an adaptive management and coordination of plans for water and coastal use, considering environmental, social and safety aspects; the indications of this directive should provide an adequate framework for the organisation of the maritime space, in combination with existing management models. Attention is also paid to different areas of coastal planning, from a tourist and cultural point of

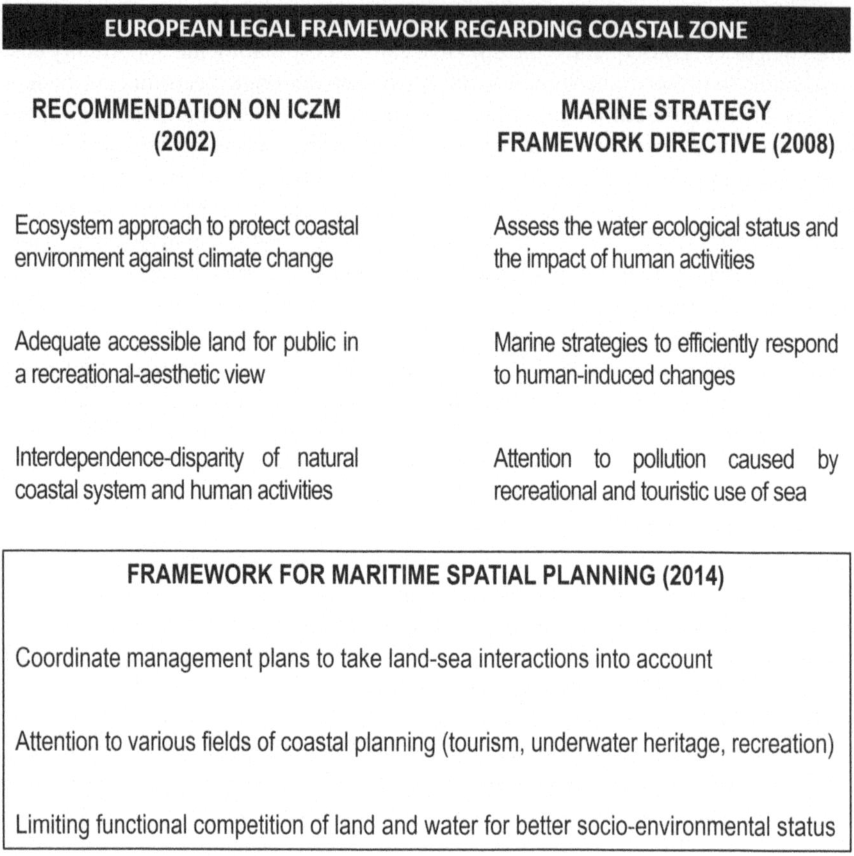

Fig. 7.1 Chart summarising the major directives related to the management and planning of the coastal and maritime space: this is relevant since the city-sea interface is both a terrestrial and a maritime territorial entity (elaboration of the author)

view, in order to limit the strong functional competition that is notoriously typical of the urban amphibious, with positive effects also on the surrounding environment: as mentioned above, in fact, marine ecosystems and resources suffer from strong anthropic pressures and natural hazards with repercussions on the quality of life, economic growth and the efficiency of coastal ecosystem services. The directive therefore recommends to carefully include these factors into plans and policies in order to have both environmental benefits and real control over the criticalities of the coastal strip with consequent improvement of socio-recreational and tourism activities. Maritime Spatial Planning can also be a useful tool to highlight crucial areas for the society: it is in fact an inclusive process that seeks to meet the needs of the coastal community in a comprehensive manner and in compliance with environmental criteria, merging, where possible, with existing systems for the management of the urban sea. Actually, the MSP tries to comprehend the specific national policies that can become an ideal basis for further maritime planning provisions, in a complementary way (Calado et al. 2022). At an international level, Directive 2014/89/EU can also be placed in the context of the United Nations 2030 Agenda for Sustainable Development, as Maritime Spatial Planning can potentially lead to the fulfillment of the Sustainable Development Goals by involving different spatial scales of maritime-coastal actions, facilitating the achievement of the SDGs at global, national and local scale (Carpenter et al. 2021). However, to date, only a limited number of European nations have integrated this guidelines into their local urban coastal management models.

7.2 Land-Sea Interactions in Maritime Spatial Planning

According to the criteria set out in the European directives, planning the city-sea interface must pay special attention to the management and spatial organisation of human activities, involving urban, natural and social sciences. This is due to the progressive dissolution of the boundaries between different littoral spaces, both in the terrestrial and maritime contexts: the coastal character appears discontinuous, which is why the management of the city-sea interface should offer a reticular approach that takes into account, alongside the environmental aspect, economic-productive issues whose benefits originate from the sea but materialise on land (for example, in the case of the fishing industry) as well as identity-cultural issues, influenced by the abovementioned gradual disappearance of traditional urban coastal frontiers (Zaucha and Gee 2019). These themes indicate the need for the active involvement of each segment of urban coastal society. In this vein, it can be stated that Maritime Spatial Planning pursues the theory of ocean governance, that is the multi-sectoral management of water and coastal space, not only from a legal perspective, but also as function of the communities that enjoy it, with regard to functional and socio-cultural aspects. So, the concept of *land-sea interactions* is included: this notion amplifies the degree of complexity in urban littoral planning but is closely related with the study of the public space of the coastal interface (Lainas 2018).

Land-sea interactions (LSI) refer to a complex phenomenon involving all the natural or anthropic processes that take place along the city-sea interface: they are related to the influences that the littoral environment has on its inhabitants and, at the same time, to the impact of human activities on the reference coastal zone. Land-sea interactions consider both negative and positive or at least acceptable effects on the marine-coastal environment, the resources offered and the human uses: LSIs can thus be defined as interactions in which land-based natural phenomena or human activities taking place on land have an influence or impact on the marine environment, the sea resources and the activities taking place in the water, and, simultaneously, as interactions in which marine natural phenomena or human activities taking place in the water have an influence or impact on the land-based environment, the coastal resources and the activities taking place on land (Iglesias-Campos et al. 2021).

Land-sea interactions can be further categorised as follows. *Interactions related to natural land-sea processes* have implications for the coastal environment and coastal socio-economic aspects that need to be identified and studied in view of their dynamic nature. The objective is to include them in planning and management processes; at the same time, they can be altered by human activities that interfere with natural processes: for this reason, the analysis of the expected impacts of land and water uses should include the assessment of their effects on natural processes within land-sea interactions and the potential consequences on natural resources and coastal ecosystem services. The *interactions between land and sea uses and activities* include the more anthropogenic component of the overall concept. It should be noted that most maritime functions require supporting installations along the coast (for instance, commercial or tourist ports, docks for recreational boating, the pipeline system connecting offshore wind farms to the mainland), as well as many functions with a mainly terrestrial basis (such as tourism and cultural uses of the coast, socio-recreational activities or bathing) expand their range of action on the water surface: such interactions can be identified and mapped, from the perspective of environmental, social and economic implications, by estimating their cumulative impacts, benefits and potential functional competition (Shipman et al. 2018). Finally, it is possible to trace a dimension of land-sea interactions that is more focused on the drafting of urban plans and management policies: the *institutional interactions*, existing between different administrative levels, can be identified between the various types of planning instruments and specific competences. The plausible convergence between maritime and land-based planning, together with the relevance of land-sea interactions, highlights the need to act in favour of an integrated system of coastal and maritime spatial planning, facilitating the transfer of knowledge from land-based planning, now consolidated in urban planning practice, to the more recent maritime management (Kidd and Ellis 2012) (Fig. 7.2).

In consideration of the variety of land-sea interactions, it is evident that they involve different types of management and planning processes that necessarily embrace areas and spaces of different extents. One of the main issues related to land-sea interactions is how wide their range of influence is, both towards the terrestrial and the water sphere. Obviously, the specific uses and the reference ecological system should suggest the dimensional extent of LSIs, in relation to the natural processes involved and the human activities occurring along the urban amphibious: these elements define the functional scope of the analysis (Asprogerakas et al. 2020). However, from the perspective of the planning process, this scope depends primarily on the spatial context of the plan to which the analysis is linked. The extent of the maritime plan, the affected portion of the shore and its features and main activities are factors that enrich and define the scope of land-sea interactions' study: natural and anthropogenic land-based processes are considered relevant to the extent that their management can have positive effects on the conditions of water areas and maritime activities, and vice versa (Schlüter et al. 2020).

In the case of coastal socio-recreational activities, this functional aspect of the littoral interface is closely connected to the natural resources offered by the coasts and their quality; at the same time, it can degrade their value: from an environmental point

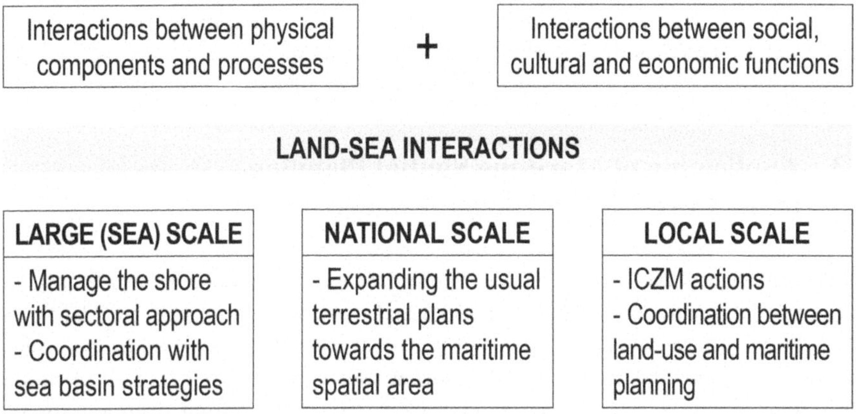

Fig. 7.2 Land-sea interactions are based on the contact between physical and coastal socio-economic systems. Depending on the scale of intervention, it is possible to calibrate specific regulatory operations; however, it should be noted that the MSP must necessarily involve both land and sea areas (elaboration of the author, based on Iglesias-Campos et al. 2021)

of view, land-sea interactions include an intensive use of coastal space that can lead to poor water and environmental quality, since tourism and recreational activities can be a source of pollution, noise and other disturbances for local species, also through the construction of buildings or infrastructures. Maritime Spatial Planning should pursue innovation in the management of new uses of the sea, so that the social-tourist function of the coastline adapts to the latest environmental standards. From a socio-economic point of view, LSIs include functional competition within coastal spaces in relation to various fields as aquaculture, energy production and port development: undeniably, this has positive effects on the economy of coastal communities but may irreversibly alter their character, occupying places that are necessary for sociality. On the other hand, tourism itself is an important economic incentive, which is why its related activities often take priority with respect to accessibility to coastal and marine areas of great landscape value: the development of tourist poles can also pose a significant threat to the water quality. Finally, from a technical point of view, land-sea interactions entail the definition of effective solutions to limit environmental pressures and ensure equal usability of the coastline, both in terms of access from the sea (for example, with pleasure boats) and in terms of accessibility to beaches and other coastal recreation sites (Ramieri et al. 2019).

In conclusion, MSP processes should take into account land-sea interactions, especially in relation to the most localised areas of coastal spatial planning, by defining the field of analysis according to the scale of intervention (which may be national, regional, sub-regional and even local), the geo-morphological characteristics of the reference coastline, the extent of LSI phenomena and their spatial and time distribution, as well as the location of maritime-coastal ecological resources. In this sense, the concept of the city-sea interface fits in as a peculiar space within which the features of the terrestrial sphere and the characteristics of the maritime space, that is now subject of studies for urban planning integration, can be contextually and coherently combined, to the benefit of the entire coastal ecosystem.

7.3 Application of Maritime Spatial Planning in the Euro-Mediterranean Context

There are twenty-two Member States along Europe's coastline: these are obliged under the MSP Directive to develop a national maritime spatial plan by 31 March 2021, with a minimum review period of ten years. At present, across Europe, states involved in plan development are at different stages of the process, with planning tools in preparation, adopted or under review. It is proposed an overview of the main states bordering the Euro-Mediterranean basin, in accordance with the target of this work, namely France, Spain, Croatia, Greece and Italy, as mentioned above, national realities that are similar from an environmental and social point of view, but profoundly different in terms of the application of maritime spatial planning.

In France, the authority responsible for maritime spatial planning is the *Ministère de la Mer*, which has the task of ensuring the coherence of strategic plans at the national level, by consulting with the *Conseil National de la Mer et des Littoraux*: the latter brings together national stakeholder representatives and reports on strategic plans to the European Commission. Moreover, the French state is the country in the Mediterranean basin that is most committed to effective coastal planning. The MSP Directive has been implemented in 2016 with the *Loi 2016-1087 du 8 août 2016 pour la reconquête de la biodiversité, de la nature et des paysages.*[1] In particular, the article no. 123 modifies the *Code de l'environnement* by introducing the notion of maritime spatial planning, considered as the process by which the state defines and organises human activities in the sea from an ecological, economic and social perspective, thus outside the spheres of defence and military security. On a national scale, the law leads to the adoption of the *Stratégie Nationale Pour la Mer et le Littoral* (Decree 2017–222 of February 23rd, 2017): its aim is supporting the coastal ecological transition process, the development of a sustainable blue economy, and the protection of good water quality and an attractive coastline. The decree aims to simplify coastal governance processes by merging several existing regional agencies into just two advisory bodies involved in the management of the French coastline from a social and ecological-environmental perspective. The strategy is implemented on a regional scale with the so-called *Documents stratégiques de façade* (DSF): these are the main tools for applying the indications of the Maritime Spatial Planning, according to the *Loi 2016-1087* and cover the four national maritime sectors along the French coast. These DSFs combine assessment procedures for environmental status and different activities along the coast (including social and bathing functions)

[1] After the Nature Protection Act of 1976 and the Landscape Protection and Enhancement Act of 1993, the French State drafted a law to restore biodiversity and landscape, protecting the natural heritage and encouraging green–blue growth.

with spatial and local planning, in order to meet the criteria of the *Stratégie Nationale Pour la Mer et le Littoral*. These strategies have become definitively operational since March 2022.

Spain established a framework for Maritime Spatial Planning by adopting the *Real Decreto 363/2017* on April 8th, 2017, which transposes Directive 2014/89/EC of the European Parliament into Spanish law. Maritime spatial management is delegated to the *Ministerio para la Transición Ecológica y el Reto Demográfico*, involving the General Directorate for Environmental Quality and Assessment during strategic and ecological impact evaluation. The planning of Spanish maritime and coastal territories is both a national and regional competence: in particular, the central government is the body responsible for international and European affairs, while regional administrations are entrusted with the management of land and water use, although national guidelines are considered in decision-making processes. Coordination between the different planning scales takes place through interministerial commissions and sector conferences. According to the *Ley de protección del medio marino* (Law 41/2010), which transposes the Marine Strategy Framework directive and regulates public domain in the maritime and terrestrial sphere, each region is in charge of producing coastal protection and monitoring strategies; in relation to the *Real Decreto 715/2012*, a *Comisión Interministerial de Estrategias Marinas* (CIEM), chaired by the Spanish Ministry of Ecological Transition, coordinates the various actions at smaller scales. In any case, in recent years the central government has worked to produce its final maritime planning strategy in line with EU guidelines, defining operational and management tools for each of the five Spanish maritime subdivisions, following a public consultation phase for the subsequent approval of the plans: indeed, the Maritime Spatial Plan (POEM) has been discussed with the population regarding the areas of high potential and the functional priorities, and then submitted to the competent authorities in order to complete the strategic environmental assessment process. The discussion also took place in an international, with Spain holding consultations on the matter with France, Portugal and Italy, with the objective of developing coastal-maritime management plans, including the action implementation and the monitoring phase, between 2022 and 2027. Actually, in February 2023, Spain finally adopted its own Maritime Spatial Planning framework, through the definitive identification of five marine areas to be managed.

As far as Croatia is concerned, there is currently no plan for the maritime space, although the entire Croatian territory is governed by various urban planning instruments, in relation to the coastline but also to the water. The *Physical Planning Act—PPA* (Act 153/13) constitutes the regulatory framework for the functional planning of the Croatian coast. Entered into force in July 2017, the purpose of this Act is to fully transpose the European guidelines for Maritime Spatial Planning in national laws. As a result of the requirements of the PPA, work has begun on the creation of a state spatial development plan to manage the entire maritime-coastal sphere, with the boundary of national territorial waters as spatial limit: in addition, spatial plans have been drawn up for protected areas managed by the national authorities, encompassing again both maritime and land space. The Ministry of Physical Planning, Construction and State Assets (MPPCSA) is the national body responsible for the planning and management of coastal land uses. At a lower scale, Croatian counties are in charge of including provisions for their marine areas in the regional level plans, regarding different types of land use: according to the PPA, the functional categories include harbours and docks, submarine cables and pipelines, protected areas, submarine archaeological sites and aquaculture areas. At the local level, municipal governments are clearly required to develop their own general urban plans, including indications to manage the uses of the marine area, if it falls within their territorial perimeter.

Similarly, Greece has not developed a maritime spatial management framework yet. However, MSP issues are faced through *Territorial Plans* included *Special Frameworks for Spatial Planning* that operate at a regional scale and cover different fields of coastal planning: so far, sectoral plans have been developed for aquaculture, recreational-tourist activities and productive sector; these tools include guidelines for coastal space and marine areas. Greece also shows some interest in renewable energy, with the 2008 *Special Framework for Renewable Energy Sources* (in the process of being updated), which defines strategic guidelines for offshore wind farms. Law 4546/2018 transposes *2014/89/EU* into Greek law, also following the updates made by Law 4759/2020: Maritime Spatial Planning instruments must follow a consultation with ministries, regions representatives and citizens; the competent authority in this field is the Ministry of Environment and Energy, in cooperation with the Ministry of Maritime Affairs and Island Policies. Today, the draft of the *National Spatial Planning Strategy for the Marine Space* is in process of finalisation for public consultation, before its final approval that will be implemented as part of the *National Recovery and Resilience Plan 'Greece 2.0'*.

To date, Italy has not officially completed or adopted a unified and binding plan for maritime space. However, Italian urban planning has incorporated the European directive through the DPCM of December 1st, 2017,[2] concerning the approval of the guidelines for the elaboration of maritime space management plans, according to which the Ministry of Infrastructure and Sustainable Mobility is the competent authority on the matter. The aim is to identify and manage the most relevant areas for land-sea interactions: among these, protected areas are included (in relation to the mutual influence that marine and terrestrial spaces have, also from a rural-productive point of view), hydrographic basins connected to the sea, stretches of coastline with great landscape-visual value in the land-sea relationship, relevant marine-coastal infrastructures and areas comprehended in port plans. Moreover, at a smaller scale, it is possible to identify maritime sub-areas, namely micro-areas of the coastal interface on which particular landscape, socio-cultural but also productive values insist. The implementation of management plans must therefore consider present and future activities related to various uses of the sea, in relation to recreation and sociality, but also to the protection of material and immaterial heritage, scientific research, transport and the exploitation of renewable energy sources. In this process, a constant

[2] This DPCM follows the principles of Decree Law 201/2016 and denotes some attention to interesting issues from the point of view of coastal sociality.

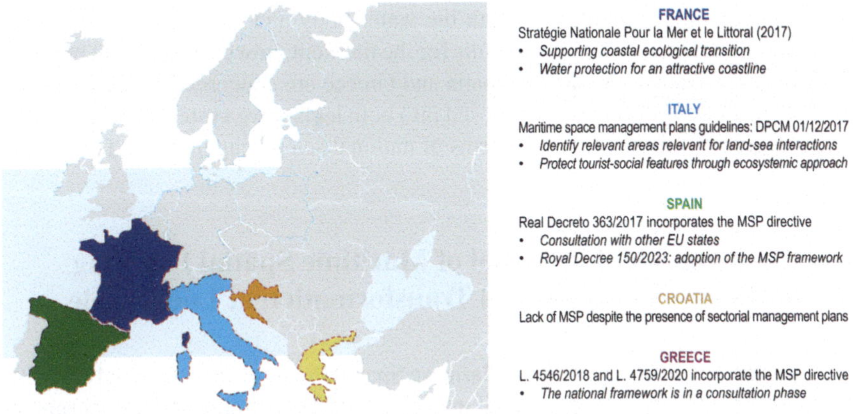

Fig. 7.3 National strategies and decrees existing in the main Euro-Mediterranean states, aimed at incorporating the MSP community directive (elaboration of the author)

inclusive cooperation between public authorities and local stakeholders is promoted[3]; at the same time, the governance of the marine space must not contrast with existing urban plans, but rather fill the gaps that coastal planning has left uncovered. It is also interesting to note that the decree encourages sustainable tourism activities, and thus sociality, along the coastal strips, supporting the adaptation of port areas to preserve historical, environmental and cultural features in the relationship between land and sea, as well as the management of sandy shores against erosion and the protection of the archaeological heritage, including the submerged one, and the traditional uses of the sea, suggesting the use of an ecosystem approach. At the regional level, it should be noted that in June 2021, proposals for future maritime spatial planning tools were submitted to the European Commission for three spatial areas, namely the *Adriatic maritime area*, the *Ionian and Central Mediterranean maritime area* and the *Tyrrhenian and Western Mediterranean maritime area*: the implementation of these plans includes public consultation phases aimed at their approval. The MSP process in Italy is currently in the final step for its definitive implementation (Fig. 7.3).

From this analysis, it thus emerges that the French State shows the more consistent progress in implementing Maritime Spatial Planning indications, while the other coastal countries denote a varying degree of attention to the issue, even though they present instruments at different scales that to some extent attempt to regulate the complex matter of coastal-maritime management. Similarly, Spain is inclined towards the further implementation of its integrated maritime strategy, especially from an international perspective: this seems interesting, as cooperation between member states is one of the cornerstones of Maritime Spatial Planning. Italy appears to be positively headed towards the final drafting of a framework indicating at several

[3] The public consultation on the future approval of MSP plans was opened on September 15th, 2022 and ended on October 31st, 2022. For further information, visit https://www.mit.gov.it/documentazione/pianificazione-dello-spazio-marittimo.

levels the planning modalities of both the water body and its related coastal area, although without having tangible results for the moment; however, there are regional initiatives in this sense. As far as Croatia and Greece are concerned, despite the fact that the EU directive has been transposed into both legislative systems, there is still a certain degree of fragmentation in terms of national sea management and planning.

7.4 Challenges and Potential of Maritime Spatial Planning for Socio-environmental Transformation at Local Scale

Even outside the European context, Maritime Spatial Planning is increasingly considered as an essential tool for the spatial management of water areas.[4] This instrument is connoted by its potential but also by challenges that appear to be spread to a more or less equal extent in the countries that have implemented it or are working to develop it, despite being in distinct biophysical, socioeconomic and political contexts: these critical issues need to be adequately addressed to ensure the suitability and long-term sustainability of Maritime Spatial Planning, in relation to the environmental status and the possibility for urban coastal society to efficiently enjoy the public coastal space and the urban sea (Frazão Santos et al. 2018). Indeed, by definition, the city-sea interface presents a blurred distinction between coastal and marine space: hence the need to think in terms of planning systems that coordinate land and sea policies and give coherence to territorial actions. However, the integration between instruments aimed at regulating the sea and instruments dedicated to the design of the coastal terrestrial sphere is not always so obvious, as these are complex spaces, within which it is difficult to precisely identify administrative competences to satisfy the social demand for accessibility to the littoral interface: as we have seen, to a certain extent, the policies of the Euro-Mediterranean basin seem to register a tendency towards the necessary planning decentralisation at the regional and local scale, although this process is still in its embryonic phase (Zaucha and Gee 2019). Within the context of the issues expressed in the first part of this research work, a synthetic but critical overview of the main problems that affect maritime spatial planning, influencing the usability of the public space of the city-sea interface, has therefore been proposed.

From a planning perspective, climate change certainly presents a global and constantly evolving challenge, since factors like global warming, sea-level rise and coastal erosion alter the distribution of assets and services in the marine-coastal ecosystem. While some maritime uses, such as renewable energy and seabed mining, are less susceptible to environmental impacts, other ones, like fisheries, aquaculture and coastal recreational functions, are generally more vulnerable: it is therefore required to plan for a constantly changing sea through increasingly flexible and

[4] According to UNESCO's *Intergovernmental Oceanographic Commission,* coastal-maritime planning tools are currently being drafted in about 70 countries worldwide, almost half of the existing coastal states, in six continents and four ocean basins (https://www.mspglobal2030.org/msp-roadmap/msp-around-the-world/).

adaptive methodological approaches (Frazão Santos et al. 2018). Within the MSP planning philosophy, the application of ecosystem-based approaches may ensure an equal spatial distribution and accessibility to coastal resources: in this sense, strategic environmental assessment proves to be an important tool for ecological implementation within maritime planning, as it estimates the potential impacts of the plan; however, as it is crucial to pursue both environmental and socio-economic objectives, it is also necessary to adequately apply the currently existing policies within the coastal interface (Ehler 2021). Indeed, the social dimension of sustainability is linked to the economic and ecological spheres through the multiple dependencies of coastal society's users on maritime resources and coastal space; consequently, the ecological decline of the sea ecosystem and the littoral sphere in general has serious consequences for coastal communities and human well-being, which underlines the importance of incorporating social justice considerations into Maritime Spatial Planning (Saunders et al. 2020; Bennett and Satterfield 2018). In particular, maritime planning should recognise and respect local arrangements and governance, as well as distinct rights and the diversity of needs expressed by social demands, pursuing equity in the distribution of benefits and defence from risks, as well as in access to resources and public spaces, with particular attention to the most vulnerable groups in coastal society; by promoting inclusion in planning and decision-making processes, MSP could be more adaptive, able to integrate social dynamics and spatial justice (Gilek et al. 2021).

From a social point of view, adequately involving stakeholders in Maritime Spatial Planning is crucial for its acceptance and adoption, because understanding the existing socio-spatial relationships, expectations and current and future interests of users allows conflicts to be reduced and economic, social and environmental objectives to be more balanced (Pomeroy and Douvere 2008). Indeed, coastal communities may suffer internal fractures when both established uses and new functions compete for the same area of sea or shore and its resources: in particular, the impact of maritime economic development may lead to conflicts of interest with coastal society users regarding recreational uses of the city-sea interface. Also for this reason, a participatory approach can improve the quality of decision-making processes, building trust and consensus in the population and leading to more durable solutions over time (Fox et al. 2013; Sayce et al. 2013). However, many factors contribute to the exclusion or lack of involvement of coastal society, including poor communication, low transparency and fragmented governance: it is not uncommon that stakeholders take part in participatory processes only in the final stages of planning drafting, rather than during the elaboration phase, when their contribution could conceivably be more incisive. Moreover, approaches like simple observations by citizens rather than more proactive approaches may limit the benefits of inclusive planning (Flannery et al. 2018; Gopnik et al. 2012). The issue of coastal community involvement is by no means secondary as it raises critical issues regarding social equity and spatial justice within the MSP.

According to this perspective, the problem of the scale of action of maritime spatial planning should also be considered. In fact, although the process of implementing water plans is usually related to specific national development objectives, it should not neglect sub-regional and local coastal contexts, taking into account the social demands of the coastal community: the goal would be to gain a systemic understanding of the integrated social and ecological components interacting at urban and micro-urban scales, since it can be argued that it is more efficient to deal with many of the socio-environmental needs at the local level (Iglesias-Campos et al. 2021). This reasoning is particularly interesting because it ties in with a thematic still not deeply investigated in the Maritime Spatial Planning field, namely the integration of coastal public space design within water plans: this is relevant because these urban areas are crucial in absorbing the impacts of contemporary environmental criticalities, through appropriate ecological planning interventions, and because they constitute a privileged point of enjoyment of the sea resource by users of the littoral interface. Places such as urban beaches, for instance, act as a hinge between the maritime system and the urban-coastal system, giving the possibility to spend leisure time in a healthy way and to exploit the water resource within congested cities: they are an expression of land-sea interactions at the human dimension and, according to their location along the urban amphibious and to their usability, they can represent a litmus test for the actual socio-recreational equity of coastal areas. In order to achieve effectiveness of Maritime Spatial Planning, it should comprehensively incorporate the complexity of the biophysical and social characteristics of coastal-maritime areas, together with relevant spatial information (Strickland-Munro et al. 2016). However, the integration of socio-environmental data is not always well established within maritime planning processes: this leads to an inadequate representation of the current social demand, which has clearly a micro-urban nature. For this reason, inaccuracies may occur in the analysis of values, perceptions, user needs, as well as the spatial and time distribution of human interactions with the coastal environment and the factors influencing it (Dalton et al. 2010; St. Martin and Hall-Arber 2008). This kind of work in the context of water plans could increase the range of decision-support tools, assist planners in their design actions and facilitate the enjoyment of ecosystem benefits for coastal community users (Le Cornu et al. 2014).

However, it should be noted that to date the legal force of maritime spatial plans is not always the same, as in some countries they are more binding, with precise indications to be followed, while in others they are more similar to guidelines that sectoral plans should follow. This brings back to the centre the need to create plans that are able to adapt to the legal, social and spatial reference context, constituting a firm support for the sustainable development of the sea and coasts, rather than a limitation to planning especially at the local scale, within which socio-environmental interactions are more interesting from the point of view of the fruition of the coast.

References

Ahlhorn F (2018) Integrated Coastal Zone Management. Status, Challenges and Prospects. Springer, Cham

Asprogerakas E, Lazoglou I, Manetos P (2020) Assessing land–sea interactions in the framework of maritime spatial planning: lessons from an ecosystem approach. Euro-Mediterranean Journal for Environmental Integration 5(18)

Bennett NJ, Satterfield T (2018) Environmental governance: A practical framework to guide design, evaluation, and analysis. Conservation Letters 11(1)

Bertram C, Rehdanz K (2013) On the environmental effectiveness of the EU Marine Strategy Framework Directive. Marine Policy 38: 25-40

Calado H, Pegorelli C, Frazão Santos C (2022) Maritime Spatial Planning and Sustainable Development. In Leal Filho W (ed) Encyclopedia of the UN Sustainable Development Goals. Springer, Cham

Carpenter A, Johansson TM, Skinner JA (eds) (2021) Sustainability in the Maritime Domain. Towards Ocean Governance and Beyond. Springer, Cham

Dalton T, Thompson R, Jin D (2010) Mapping human dimensions in marine spatial planning and management: an example from Narragansett Bay, Rhode Island. Marine Policy 34(2): 309–314

Douvere F, Ehler CN (2011) The importance of monitoring and evaluation in adaptive maritime spatial planning. Journal of Coastal Conservation 15: 305–311

Ehler C (2021) Two decades of progress in Marine Spatial Planning. Marine Policy 132

Flannery W, Healy N, Luna M (2018) Exclusion and non-participation in Marine Spatial Planning. Marine Policy 88: 32–40

Fox E, Poncelet E, Connor D, Vasques J, Ugoretz J, McCreary S, Monie D, Harty M, Gleason M (2013) Adapting stakeholder processes to region-specific challenges in marine protected area planning. Ocean & Coastal Management 74: 24–33

Frazão Santos C, Ehler CN, Agardy T, Andrade F, Orbach MK, Crowder LB (2018) Marine spatial planning. In Sheppard C (ed) World Seas: An Environmental Evaluation, Volume III: Ecological Issues and Environmental Impacts. Elsevier, Amsterdam, p. 571–592

Gilek M, Armoskaite A, Gee K, Saunders F, Tafon R, Zaucha J (2021) In search of social sustainability in marine spatial planning: A review of scientific literature published 2005–2020. Ocean & Coastal Management 208: 1–14

Gopnik M, Fieseler C, Cantral L, McClellan K, Pendleton L, Crowder L (2012) Coming to the table: Early stakeholder engagement in marine spatial planning. Marine Policy 36(5): 1139–1149

Iglesias-Campos A, Rubeck J, Sanmiguel-Esteban D, Schwarz G (eds) (2021) MSPglobal. International Guide on Marine/ Maritime Spatial Planning. UNESCO-IOC/European Commission, Paris

Kidd S, Ellis G (2012) From the Land to Sea and Back Again? Using Terrestrial Planning to Understand the Process of Marine Spatial Planning. Journal of Environmental Policy & Planning 14(1): 49–66

Lainas I (2018) Land-sea interactions and maritime spatial planning guidelines in the context of European Union. The case of Greece. RELAND: International Journal of Real Estate & Land Planning 1: 365–376

Le Cornu E, Kittinger JN, Koehn JZ, Finkbeiner EM, Crowder LB (2014) Current practice and future prospects for social data in coastal and ocean planning. Conservation Biology 28(4): 902–911

Moksness E, Dahl E, Støttrup J (eds.) (2013) Global Challenges in Integrated Coastal Zone Management. Wiley, Hoboken

Morales JA (2022) Mitigation, Coastal Policies and Integrated Coastal Zone Management. In Morales JA, Coastal Geology. Springer, Cham, p. 447–455

Pomeroy R, Douvere F (2008) The engagement of stakeholders in the marine spatial planning process. Marine Policy 32(5): 816-822

Ramieri E, Bocci M, Markovic M (2019) Land Sea Interactions in the framework of ICZM and MSP. Priority Actions Programme Regional Activity Centre (PAP/RAC)

Saunders F, Gilek M, Ikauniece A, Voma Tafon R, Gee K, Zaucha J (2020) Theorizing Social Sustainability and Justice in Marine Spatial Planning: Democracy, Diversity, and Equity. Sustainability 12(6)

Sayce K, Shuman C, Connor D, Reisewitz A, Pope E, Miller-Henson M, Poncelet E, Monie D, Owens B (2013) Beyond traditional stakeholder engagement: public participation roles in California's statewide marine protected area planning process. Ocean & Coastal Management 74: 57–66

Schlüter A, Van Assche K, Hornidge AK, Văidianu N (2020) Land-sea interactions and coastal development: An evolutionary governance perspective. Marine Policy 112

Shipman B, Roberts H, Dworak T, Zamparutti T, Krüger I, Veidemane K, Mashkina O, Parrod C, Ceresil E, Moarcas A, Oulès L (2018) Land Sea Interactions in Maritime Spatial Planning. EU MSP Platform

St. Martin K, Hall-Arber M (2008) The missing layer: Geo-technologies, communities, and implications for marine spatial planning. Marine Policy 32(5): 779–786

Strickland-Munro J, Kobryn H, Brown G, Moore SA (2016) Marine spatial planning for the future: Using Public Participation GIS (PPGIS) to inform the human dimension for large marine parks. Marine Policy 73: 15–26

Zaucha J, Gee K (eds) (2019) Maritime Spatial Planning. Past, present, future. Palgrave Macmillan, Cham

Chapter 8
Geo-spatial Approach for Studying the Features of the City-Sea Interface

Abstract Governing urban coastal areas involves grasping their intricate physical and functional structure. Such spaces rely on various interconnected variables, including physical, natural, and socio-economic factors. For this reason, it is relevant to apply a multidisciplinary approach across different scales and timeframes. In particular, the chapter explores a geo-spatial approach in order to develop a spatial data model for georeferencing information concerning the city-sea interface. This tool could facilitate the representation of spatial components as geometric objects with specific features. Its utility lies in accurately depicting littoral components, supporting the general understanding of the urban amphibious and its management.

Keywords Spatial data model · Geo-spatial approach · Spatial data · GIS · Coastal analysis · Socio-recreational uses · Littoral ecological assets

8.1 The Complexity of the Contemporary Coastal Landscape

It is usual to think of coastlines as mere material objects, ignoring the degree of complexity that lies within them. The composition of coastal elements is extremely articulated, as it is possible to identify recreational spaces for sociality, but at the same time natural amenities, wetlands that are composed of specific vegetation, stratified ecosystems that are home to numerous animal species, as well as vulnerable zones susceptible to different types of risk. Certainly, this articulation is not static but changes over time. For this reason, it is good to think of coastal areas as processes, which are not fixed but continuously reinvent themselves or are subject to external transformations. Continuous coastal evolution involves the interaction between solid materials (like sand or rocks), chemical elements (such as salinity or the composition of the seas), energy flows of physical, biological and even anthropic nature, ecological activities and, of course, deep-rooted human activities. It is therefore logical to expect accelerated responses from the environment, in relation to the complexity of the coastal site: this is due to the rapid aggravation of climate change, anthropogenic

I. Pistone, *Urban Coasts in Socio-ecological Transition*,
https://doi.org/10.1007/978-3-031-70783-4_8

modifications but also to an inherent criticality underlying the use of littoral space, especially within the urban system, in relationship with the presence of differentiated interests and needs in a very limited space (Donelson Wright et al. 2019) (Fig. 8.1).

The contemporary coastal landscape is thus presented as the result of complex interactions between natural processes and human activities. As previously stated, urban growth and the increase of natural hazards related to the maritime environment create new challenges for coastal management: in order to study the city-sea interface, the quantification, modelling and visualisation of short-term evolution in relation to environmental and anthropogenic impacts may be useful. Identifying sensitive areas and neuralgic points for sustainable transformation, as well as mapping socio-ecological vulnerabilities, are relevant steps for responsible urban coastal planning and management (Hardin et al. 2014). The coastal strip, especially in the theoretical context of the city-sea interface, is in fact a space with relatively reduced width but characterised by considerable length and strong pressures, since it is the site of multiple environmental, social and economic processes and criticalities, particularly in recent years, during which an increasing concentration of uses of the sea and urban shorelines can be registered (Gourmelon and Robin 2005): this situation translates into functional conflicts and limited access to resources, which are difficult to

Fig. 8.1 Contemporary coastal landscape is affected by multiple stakes and functions, as well as growing risks from the sea, which undermine its integrity and usability. In the photo, coastal urbanisation in the zone of Los Angeles (from Wikimedia Commons, licensed under CC BY-SA 4.0)

manage due to mono-disciplinary approaches for critical coastal issues, to an unclear definition of administrative and legal competences and to legal regulations that are juxtaposed but not really integrated with each other. It is therefore not surprising that littoral zones represent a vulnerable space where meteorological events and marine impacts combine with pressures from competing human activities, amplifying risks of natural and anthropogenic origin.

As is now well known, the percentage of inhabitants concentrated along urban coastlines is steadily increasing worldwide, with very high rates of development in the Euro-Mediterranean area: among the numerous other constraints on sustainable development, one of the consequences is a tangible reduction of the amount of natural urban areas. Large cities undeniably trigger socio-economic progress, but the downside is massive urbanisation that leads to air and water pollution, inadequate drainage of rainwater and limited green spaces and land available for the various activities of communities. The result is the progressive homologation of urban and coastal landscapes and the creation of constraints on the socio-environmental safety, the city ecosystem and the aesthetic and recreational value of urban agglomerations: to ensure urban sustainability, it is therefore necessary to develop and implement innovative and integrated solutions that address the specific social, economic and environmental issues of these places, before the complexities of the contemporary city (and its coastal areas) become so articulated that the damage caused so far will be irreparable (Bougdah et al. 2020). In the coastal environment, it seems that measuring the impact of human activities constitutes both one of the main challenges of management and planning at different scales and one of the major difficulties encountered by the multidisciplinary study of the coastal zone in recent decades (Caldwell and Hartge 2013). This critical issue is linked to the complexity of analysing the role of coastal societies within coastal governance processes due to its constantly changing nature, the impossibility of applying rigid and deterministic methods to this type of study and also the difficulty in finding structured information describing the anthropic impact on the environment. In this context, understanding nature-society interactions within the city-sea interface again leads to the need to define multidisciplinary analytical-operational frameworks for the spatialisation of coastal dynamics (Kumar et al. 2019).

The complexity of these spaces can be reinterpreted through the collection and study of data about the physical structure of the city-sea interface and its spatial and functional organisation. The digital reconstruction of the territorial context of the urban amphibious can provide interesting insights in order to better understand the conditions within which the public littoral space is most usable by the users of the coastal society, or on the contrary which are the physical-functional constraints that prevent its valid and equal fruition. Thus, spatial-geographical analysis is a useful approach to investigate the issues enunciated by the research question in relation to social, environmental and geophysical aspects (Donelson Wright et al. 2019). Increasingly, however, scholars are also relying on research based on non-expert knowledge, actively involving society through different methods. Indeed, it is possible to achieve high-quality results in the decoding of the coastal territorial system through minimal efforts on the part of users, which, in combination with research work, can produce vast datasets that are extremely useful in the analysis, monitoring and management phases: such approaches could help in better describing the landscape composition as a function of its spatial configuration, understanding which factors influence its interpretation and quantification in analytical terms. The increased availability of spatial data, particularly in the last two decades, has shown new opportunities for landscape development, management and transformation. Coastal research should therefore investigate the interactions between different socio-ecological processes in order to describe how littoral urban zones may change over time, also on the basis of comparisons between related spatial systems (Gergel and Turner 2017).

To shed light on the complexity that hovers over the coastal territory and in particular over the city-sea interface, it can be a useful support to evaluate and possibly implement alternative strategies that favour the understanding of the public space: in particular, in this research field, the geo-spatial approach can highlight the main characteristics of the coast in metric and quantitative form, as a basis of information for further comparison with other littoral systems. The resulting geo-spatial model could describe the coastal thematic area analysed in this work, but at the same time it could foster reflections on its main attributes and criticalities and possible areas of conflict from a functional, environmental and management perspective.

8.2 The Concept of Spatial Data Model

On the basis of this reasoning, it is proposed the realisation of a *spatial data model*, a tool that allows the georeferencing of information related to a geographical space, such as the city-sea interface. Its usefulness lies in its ability to represent the components of space in the form of geometric objects with precise characteristics related to the reference research (Neteler and Mitasova 2004): georeferenced data in fact include a spatial component (geometric or graphic), which describes the position or spatial distribution of a phenomenon or function, and attributes that describe its intrinsic properties. In terms of representation, the spatial component can be expressed through geometric objects, whose characteristics are defined as lines, points and areas in relation to their geographical coordinates, defining a vector data model. Depending on the scale, the representation of geographic-functional features can obviously change: consequently, the urban coastal system can be represented as a point or an area, according to the specific needs (Balram and Boxall 2019). Such georeferenced data can be translated into layers for the composition of thematic maps: each feature or attribute at the base of the coastal interface can be combined with other information to carry out new elaborations and produce analyses of the different functional and spatial relationships: in this way, it is possible to decode and visualise the various processes underlying the communities that inhabit the many zones of contemporary city, such as accessibility, specific uses, demographic growth, but also aspects that are more related to the physical-environmental system and the evolution of the ecosystem (Sharma 2021).

This functionality of the spatial data model can be used to address critical issues in today's spatial planning and management: its application from a socio-economic perspective can be useful to identify solutions to the problem of local accessibility, to locate the most relevant points of the urban coastline within a certain range, to find the optimal position for organising uses and functions in a part of the city, to assist in the design of community services in relation to the demographic composition of the coastal interface, spatially identifying the expectations of society and the peculiar characteristics of the urban system, essentially managing urban growth (Ivan et al. 2017). This geo-spatial approach can also play an important role in natural resource conservation and disaster management: it can be applied for actions such as identification and prevention of coastal erosion risk, resource organisation, ecosystem analysis and modelling, environmental planning, flood protection and management, but also to map human activities that may generate pollution (Bougdah et al. 2020).

It is indeed appropriate to investigate how to manage a complex space such as the urban amphibious, fully understanding its physical-functional composition: its functioning depends on numerous physical, natural and socio-economic variables that interact with each other at different scales and over a more or less broad time range, requiring a multidisciplinary approach. The coastal sphere of large contemporary cities thus appears as a space that is sensitive to a variety of parameters to be interpreted, also in relation to the interests of coastal users, whose frequent conflicts may be resolved also thanks to a geographic-spatial study (Gourmelon 2003). Geographical information systems are, for the most part, developed to study phenomena according to an approach that tends to be mostly static, although it is possible to integrate spatial analysis with time variables, elaborating comparisons and maps in a given chronological interval. Combining the spatial data model with other types of survey processes or methodological approaches gives access to the dynamics underlying urban development (Bédard et al. 2007). The analysis of spatial data[1] requires a reference framework to systematise the studied phenomena and elements. In this sense, it is possible to identify a twofold approach to geographical representation in research[2]: the discrete method favours the study of discretised elements that fill the space, while the continuous method focuses on the study of space, in this case the coastal space, as covered by surface elements that characterise it, also from a functional perspective (Haining 2003). The classification of spatial data according to the type of spatial conception (together with the choice of an appropriate scale of analysis) is a necessary step in establishing the appropriate methodologies and techniques to answer the research question.

[1] Spatial data refer, directly or indirectly, to a specific geographic area (which is why they are sometimes referred to as geospatial data): they may represent a physical object in a geographical coordinate system, but they are also evidently useful in the study of how different variables may affect individuals, communities and the environment (Kumar et al. 2019).

[2] In statistics, a quantitative variable is said to be discrete when it can assume a finite or numerable set of values; conversely, a variable is described as continuous when it can theoretically assume all the values included in a real interval (Bramanti 1997): consequently, an object of study (such as space and its elements) can be considered continuous if it contains an indefinite number of elements arranged contiguously, or discrete if it is composed by isolated, non-contiguous elements.

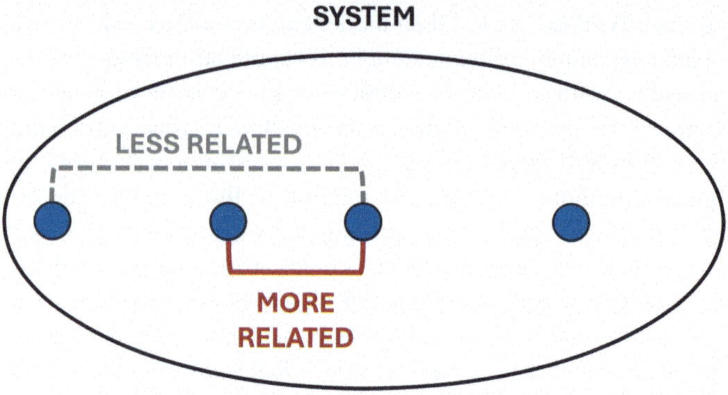

Fig. 8.2 The system defined by Tobler's third law of geography, according to which the elements closest to each other have more intense mutual connections (elaboration of the author)

In this context, it is possible to identify four different types of spatial data, broadly speaking: *point pattern data* indicate a set of information that is organised in a punctual manner within the study area, according to the positions in which interesting events occur (such as specific uses or the impact of particular phenomena); *field data* refer to variables that are conceptually continuous but are sampled and studied according to fixed criteria that establish an observation interval, like geostatistical data; *area data* associate the analysed information with a certain number of areal entities, forming regular grids (as in the case of remote sensing images) or surfaces with more or less regular shapes (for instance, neighbourhoods, districts or census sections, but also functional areas); finally, *spatial interaction data* refer to measurements of phenomena or spatial-functional entities that occur in relation to a precise position or certain areas (Fischer and Wang 2011). The study of spatial data is related to the principle that the correlation between variables in geographically or functionally close locations tends to be stronger than the relationships between variables in distant locations: there is thus an inversely proportional relation between the spatial-functional distance of the elements of analysis and their similarity.[3] Obviously, this proposition admits exceptions: it is possible for variables that are very close to each other to be contradictory, just as very distant elements may show great affinity; again, there are values of certain variables that are independent of their location (Anselin and Rey 2010). The usefulness of this geo-spatial approach thus emerges in the study of elements belonging to the same spatial system (such as the city-sea interface), but also in the comparison of spatial entities that are geographically distant but functionally related (Fig. 8.2).

From a methodological point of view, it is necessary to define the motivation of spatial research and the perspective towards which it tends.

[3] Tobler's first law of geography (Tobler 1970) declares that everything is related to everything else, but things close to each other tend to have more intense relationships than things far away from each other.

The purpose of the study in fact constitutes the interface between the dimensional structure of the investigation and the socio-environmental context that motivates it: the aims can obviously be oriented towards theoretical analysis, but also towards operational and institutional-decisional support. For this reason, it is important to correlate the intentionality of the research with the production of expected representations and outputs, which will have to be designed to satisfy the initial purpose. It will also be relevant to distinguish selection criteria for the elements to be analysed, filtering what does not fit into the category of the object of study, but also categorising the hierarchy of information to be collected. In particular, the hierarchical structure is composed by layers that allow for the gradual decoding of the space of the coastal interface: spatial and time aspects are then combined with a variety of attributes that semantically reorder the territory, giving back a comprehensible and coherent picture, suitable for in-depth research (Tambassi 2019). The elaboration of thematic maps of the coastal areas of contemporary cities allows to delve into urban forms, management processes and morphological-environmental evolution, providing an important background to different branches of coastal knowledge, starting from natural sciences and ending with legislation, environmental engineering, architecture and all those branches that flow into urban planning practice and theory. The use of these outputs can certainly have different purposes, but they converge in the realisation of a platform that allows the management of data, visualising the various physical and functional information, interpreting and analysing the current and past conditions with a view to improving the quality of future coastal conditions. Hence, the realisation of a spatial data model may allow to monitor and evaluate changes in the city-sea interface, digitally organising littoral resources for coastal space management and planning (Pérez-Alberti et al. 2021).

8.3 Structure of a Spatial Data Model for the City-Sea Interface

The main objective of the proposed framework for a spatial data model dedicated to the urban amphibious is the analysis of the social and ecological elements of the city-sea interface, defining its morphological and functional characteristics. This can shed light on the socio-recreational potential of these places: to this end, it appears relevant to study which social uses directly related to the sea are practised along the urban shoreline, as well as the actual presence of coastal ecosystem services, especially related to the psychophysical well-being of citizens and the enjoyment of coastal cultural, landscape and natural amenities; this is linked to the analysis of the current morphological structure of urban shorelines as well as the formulation of reasoned hypotheses regarding its possible future evolution. During the last 50 years, major transformation and regeneration of urban coastlines have occurred, as stated above: this makes interesting to analyse the littoral interface from a spatial point of view. The model therefore involves a diagnosis on a city scale aimed at studying the types of public and semi-public space along the coastal interface, the accessibility to the coast, the endowment of ecosystem services; at the same time, it gives the possibility of investigating some specific sites of the reference coastal interface, delving into the theme of fruition of the sea, visibility along the coast and socio-recreational coastal uses. The hypothesised structure can thus be articulated according to groups of layers, schematised as follows (Gourmelon and Robin 2005).

- *Base maps* will include aerial photographs, topographic maps, three-dimensional models and georeferenced maps within a given time frame or in a precise historic moment. This information allows the current coastline to be compared with spatial configurations assumed in the past, but also allows for the updating and mapping of data in the next categories.
- The *geo-physical structure layer* will collect data on the physical structure of the coast and the elements that modify its profile, whether they are anthropic or natural: in particular, reference is made to the coastline (understood as the perimeter of the urban shoreline) and to the anthropic elements designed for preserving the coastal interface from environmental risks or for social and recreational purposes (such as artificial rocks, piers, docks). The objective of this layer is to give shape to the outer border of the maritime city edge, in order to precisely locate land and sea activities, but also to relate the functional uses and environmental components of the interface with respect to the sea front; moreover, it allows to know the extent of the defence works against the maritime risk, a crucial element for studying the use of the urban coast for recreational purposes.
- *Natural environment layer* refers to the green spaces located along the city-sea interface, whose ecological and environmental values counterbalance the negative impacts of the urban fabric: it is possible to identify vegetated areas characterised by different types of urban greenery (including private gardens, urban parks of various sizes, aligned trees, but also abandoned or degraded natural areas and urban rural spaces, as well as generic urban forestation) and protected areas, both in the marine and terrestrial spheres. This layer allows to analyse the available surface for green–blue implementation of various types, also studying the amount of vacant open areas: this data can be related to the percentage of sealed soil within the city-sea interface but can also provide information about the presence of more or less extensive elements to be included in the network of coastal green–blue infrastructure.
- Within the *built environment layer,* artificial and anthropic elements characterising the city-sea interface fall into two macro-categories, namely buildings and port areas. The former will be classified according to their function, in particular with reference to residential and tourist uses, nautical clubs, public facilities, production plants, possible military bases, but also buildings with historical-cultural value belonging to the terrestrial or submerged heritage and disused or abandoned buildings. The second macro-category will investigate all types of port infrastructure, considering harbours, private piers, commercial, tourist or military ports. Although the primary purpose of the present spatial model is to study the socio-recreational uses and environmental aspects of the city-sea interface, it is

nevertheless important to understand the composition of all parts of the urban coastline.

- The *recreational uses layer* identifies areas of the urban amphibious intended for bathing, both public and private, but also artificial platforms used as solariums; public and private spaces dedicated to outdoor sporting activities, like playgrounds or gymnastics areas, sports fields or outdoor swimming pools are identified too. The goal of this layer is to identify punctually socio-recreational activities related to the urban sea: indeed, the amount of public spaces along the littoral zone is at the core of the social demand expressed by the urban coastal society, regarding the possibility of fruition of these places, in terms of direct accessibility (free access, access with a ticket, private access, access physically or legally denied) but also indirect accessibility, in relation with potential constraints to visibility (Nikkoli 2018).
- The last layer analysed in this framework is related to *soft mobility and accessibility* and includes everything that concerns the possibility of reaching the coast by road and rail or by sustainable means of transportation. In fact, it traces driveways, private routes, railway or tram lines along the coast and areas intended for parking cars or motorbikes (considering public, paid or private parking areas); furthermore, this layer measures pedestrian areas, understood as road sections where vehicular traffic is forbidden, public squares, promenades and all those spaces legally dedicated to pedestrians and cyclists in a clearly designed manner. Access to the public space within the city-sea interface is clearly connected to the transport infrastructure and measuring its extent can provide useful insights about how much of the urban shore is actually usable by its citizens.

This framework is certainly flexible, offering the possibility of being implemented or readapted according to the study objective, while focusing strongly on the aspects of sociality and the environmental dimension along the city-sea interface at a local level, in order to provide a database for developing policy decisions and coastal plans at a smaller scale. Specifically, the expected outputs concern the production of maps and measurements of the different coastal elements that actually compose the littoral interface, gaining an in-depth understanding of its composition by localising the main current recreational uses that take place there, the amount of vegetated coastal spaces, the morphological composition of the present-day urban coastline and the percentage of equally accessible areas along the city shore; to this, information concerning the environmental status of the urban amphibious and the elements intended for protection from climate risks can be integrated (Table 8.1).

Table 8.1 Summary diagram of the proposed structure of the spatial data model for the city-sea interface

Spatial data model				
Layers	Elements	Description	Possible attributes	Geometry
0. Basic layers	Aerial photographs	Most recent and precise image source to analyse the coastal interface	/	/
	Georeferenced maps	Base maps describing landscape elements (topography, buildings, roads, coastline, etc.)	/	Line
1. Geo-physical structure	Coastlines	Natural or artificial perimeter of the urban coast	• Natural coast • Artificial coast	Line
	Coastal defence works	Artificial elements built to preserve the coast from the sea or for recreational purpose, modifying its perimeter	• Dykes • Piers • Breakwaters • Artificial rocks	Polygon
2. Natural environment	Vegetated areas	Areas with various types of vegetation, including spontaneous ones	• Private gardens • Public gardens or parks • Urban forestation • Aligned trees • Abandoned green areas	Polygon
	Protected areas	Terrestrial and marine natural parks within the city-sea interface	• Land areas • Marine areas	Polygon

(continued)

Table 8.1 (continued)

Spatial data model				
Layers	Elements	Description	Possible attributes	Geometry
3. Built environment	Buildings	Constructions used for residential, economic, productive or sporting functions	• Residential buildings • Accommodation buildings • Nautical clubs facilities • Historical-cultural buildings • Public buildings • Productive buildings • Military buildings • Disused buildings	Polygon
	Port areas	Areas dedicated to commercial, productive, leisure and military port infrastructures	• Leisure public ports • Leisure private ports • Commercial ports • Tourist ports • Military ports	Polygon
4. Recreational uses	Bathing areas	Public or private areas of the coast used for bathing	• Public bathing areas • Private bathing areas • Artificial platforms	Polygon
	Outdoor sport facilities	Public or private areas for outdoor sports and training or outdoor swimming pools	• Outdoor playgrounds • Outdoor swimming pools	Polygon
	Boat storage	Places where private boat are stored along the shore	/	Polygon
5. Soft mobility and accessibility	Pedestrian areas	Road section along the coast for pedestrians or cyclists	• Squares • Pedestrian areas • Sea promenades • Sidewalks • Cycle lanes	Polygon

(continued)

Table 8.1 (continued)

Spatial data model				
Layers	Elements	Description	Possible attributes	Geometry
	Railways, tramways	Electric or traditional rail connections along the coast	• Railways • Tramways	Line
	Car roads	Streets where cars are allowed within the coastal interface	• Public roads • Private roads	Line
	Parking areas	Places where parking of vehicles is permitted	• Public car parking • Public motorbike parking • Paid parking • Private parking	Polygon

Elaboration of the author in collaboration with CNRS ESPACE—UMR 7300

References

Anselin L, Rey S (2010) Perspectives on Spatial Data Analysis. Springer, Cham

Balram S, Boxall J (eds) (2019) GIScience Teaching and Learning Perspectives. Springer, Cham

Bédard Y, Bernier E, Badard T, Chrisman NR (2007) Research in geographic information systems, data management and dissemination, and new geospatial technologies. Geomatica 61(3): 299–314

Bougdah H, Versaci A, Sotoca A, Trapani F, Migliore M, Clark N (eds) (2020) Urban and Transit Planning. A Culmination of Selected Research Papers from IEREK Conferences on Urban Planning, Architecture and Green Urbanism, Italy and Netherlands. Springer, Cham

Bramanti M (1997) Calcolo delle probabilità e statistica. Società Editrice Esculapio, Bologna

Caldwell MR, Hartge EH (2013) Coastal Issues. In Garfin G, Jardine A, Merideth R, Black M, LeRoy S (eds) Assessment of Climate Change in the Southwest United States: A Report Prepared for the National Climate Assessment. Island Press, Washington, DC, p. 168–196

Donelson Wright L, Syvitski JPM, Reid Nichols C (2019) Coastal Complexity and Predictions of Change. In Donelson Wright L, Reid Nichols C (eds) Tomorrow's Coasts: Complex and Impermanent. Springer, Cham, p. 3–23

Fischer MM, Wang J (2011) Spatial Data Analysis. Springer, Cham

Gergel SE, Turner MG (eds) (2017) Learning Landscape Ecology. A Practical Guide to Concepts and Techniques. Springer, Cham

Gourmelon F (2003) La contribution des SIG à la connaissance et à la gestion de l'environnement littoral. Université de Bretagne occidentale, Brest

Gourmelon F, Robin M (eds) (2005) SIG et littoral. Lavoisier, Cachan

Haining R (2003) Spatial Data Analysis: Theory and Practice. Cambridge University Press, Cambridge

Hardin E, Mitasova H, Tateosian L, Overton M (2014) GIS-based Analysis of Coastal Lidar Time-Series. Springer, Cham

Ivan I, Singleton A, Horák J, Inspektor T (eds) (2017), The Rise of Big Spatial Data. Springer, Cham

Kumar D, Singh RB, Kaur R (2019) Spatial Information Technology for Sustainable Development Goals. Springer, Cham

Neteler M, Mitasova H (2004) Open-Source GIS. A GRASS GIS Approach. Kluwer Academic Publishers, Amsterdam

Nikkoli A (2018) Les rives lacustres comme espaces publics: essai de cartographie. L'exemple du lac d'Annecy. Mappemonde 123

Pérez-Alberti A, Pires A, Chaminé HI (2021) Shoreline and Coastal Terrain Mapping. In Finkl CW (ed) Encyclopedia of Earth Sciences Series. Springer, Cham, p. 1558–1571

Sharma P (ed) (2021) Geospatial Technology and Smart Cities. ICT, Geoscience Modeling, GIS and Remote Sensing. Springer, Cham

Tambassi T (2019) The philosophy of GIS. Springer, Cham

Tobler WR (1970) A Computer Movie Simulating Urban Growth in the Detroit Region. Economic Geography 46: 234–240

Chapter 9
Psycho-social Approach to Analyse Coastal Social Perception

Abstract Considering the complexity of coastal territory and the high expectations of its users, it becomes essential to understand the perceptive component underlying the liveability of coastal open spaces. These places have indeed a high impact on coastal society and are essential for sustainable spatial design. Thus, the implementation of socio-recreational practices along the city-sea interface should deal with coastal-environmental protection and the enhancement of citizens' well-being. It is here presented a psycho-social approach directed towards quantitative research in order to verify assumptions through field surveys and questionnaires, translating concepts into measurable variables.

Keywords Quantitative analysis · Questionnaire · Psycho-social approach · Social representation · Concept of representativeness · Urban coastal society

9.1 The Social Perception of the City-Sea Interface

The contemporary territory is characterised by a singular complexity of signs that anthropic actions imprint on the forms of nature: even in the case of areas that are still scarcely urbanised, a place remains uncontaminated only until it is not subject to the perception of an observer, who organises forms and spaces in a semiotic mental model filtered through previous experiences. Indeed, the concept of landscape is in any case a mental operation of man (Luginbühl 2017). According to the European Landscape Convention (2000), the term landscape refers to people's perception of a certain part of a territory whose character derives from the interrelations between environmental and human factors that the observer is able to recognise, incorporating visual, emotional, socio-cultural and psychological components. It is possible to connect the social perception of the landscape, including the coastal landscape, to different scales: today, the participation and involvement of users in the planning and management of their territory emphasises the social perception on a local scale, which considers the attachment to place, social relations and the individual experience of citizens. The perceptive dimension is a key element of landscape in

I. Pistone, *Urban Coasts in Socio-ecological Transition*,
https://doi.org/10.1007/978-3-031-70783-4_9

relation to territory. However, it is a concept that is difficult to quantify: investigating social perception means first of all exploring the different values of the landscape, which include naturalness and recreational enjoyment. Such a study also refers to the sources of social perception, linked to expert and non-expert users, to psychological and mental constructs of perception (in psychophysical and cognitive terms), to quantitative or qualitative analytical approaches and to possible concepts for measuring perception, such as the physical qualities of a certain place, the preference declared by users or the landscape beauty (Cassatella 2011): it is clear that visual perception alone does not determine users' preferences, yet it plays an important role in defining a community's need for sociality and well-being, also through hypotheses of planning and design intervention.

With regard to the study of urban shorelines, it has been stated that open spaces undeniably bring benefits to human health, positively influencing the psycho-physical well-being of users: many studies report that frequenting such areas can restore psychological status by replenishing users' cognitive resources, induce a positive emotional balance and even have an influence on the hormonal physiology related to the nervous system, reducing the average stress level (Bratman et al. 2021). According to psycho-evolutionary theory, humans have in fact evolved to recover more quickly from psychologically, physiologically and socially traumatic situations in valuable environments, free of threats and rich in resources and services, rather than in oppressive, highly urbanised contexts. Furthermore, the regenerative capacity of open spaces and blue-green areas increases when the perception of these places is reassuring and appealing for users, compatible with their lifestyle and needs. It has been demonstrated that components that increase or decrease the naturalness of public space, such as urban vegetation or the characteristics of urban parks, can actually influence how society perceives places, expanding the potential for environmental and psychological restoration (Georgiou et al. 2021).

Alongside the natural and urban components that define the public space of the coastal interface, the presence of other people can be an additional element that may alter the perception of places and the potential for psychological restoration: the presence of users considered non-threatening or overcrowded places can in fact impact the degree of perceived safety and well-being (Collado et al. 2017). Along the city-sea interface, however, there are public spaces of different size and degree of fruition: it is well established that the urban shore is perceived by coastal society as an important opportunity for socio-ecological transformation. In particular, spaces such as urban beaches consistently promote the psychological restoration of the coastal community, just as the juxtaposition of different types of green–blue areas along the city shoreline can have different impacts on social perception, also from a planning and management perspective. The excessive fragmentation of naturalistic areas of the urban coast due to anthropic elements, such as port areas or piers, may incredibly reduce or even completely erase the users' perception of well-being; just as spending leisure time in spaces with great social value like urban beaches can have positive effects on the psycho-physical status of the users, in the same way the presence or lack of services can alter, at the micro-urban scale, the potential for psychological restoration of these places along the city-sea interface. There are environmental variations in the way the public and open space of the urban coastline is perceived and in the well-being that these places provide for the citizens; moreover, the different individual experiences can modify the social perception of the urban amphibious (Subiza-Pérez et al. 2020). An example may be the change in the fruition of public places following the COVID-19 epidemic, a phenomenon that has expanded the need for proximity open areas, focusing markedly on social distancing, health protection and availability of personal space (Severin et al. 2021; Moreno 2020).

Regarding the perception of comfort and well-being that users may have in different parts of contemporary coastal areas, this aspect can be influenced by several factors. Specifically, the social perception of the coast may vary according to the naturalness of the place: generally, users prefer urban beaches and dune areas to markedly anthropic places such as jetties or piers, while urban parks and zones with a historical-cultural character possess an average value between urban naturalness and urbanised spaces. In essence, the perception of psycho-social well-being is less pronounced in areas evidently defined by human action and strongly artificialised, although the presence of coastal services favours the use of more ecologically valuable spaces (Hooyberg et al. 2022). This clearly highlights the importance of the socio-environmental component for the well-being of urban coastal society: considering this element as an aim for littoral planning and policy could lead to interventions that maximise the benefits provided by the city-sea interface to its users.

Understanding the impact of coastal open spaces on social perception is a fundamental prerequisite for sustainable spatial design and the development of new socio-recreational practices and functions along the city-sea interface in accordance with the protection of the coastal environment, with positive impacts on the psycho-physical well-being of citizens. However, it is compulsory to analyse how the multi-sensory and functional experiences of urban coastal society influence the psycho-social effects of the city shore and how these effects may differ between segments of littoral communities with different demographic and socio-economic backgrounds: this could be influenced by climate risks, as discussed above, but also by the overcrowding of coastal public places and the possible lack of accessibility, as well as the physical and aesthetic conditions of these areas (Hipp and Ogunseitan 2011). The short-term and long-term psycho-social benefits of using the littoral areas in the city could also be quantified from an economic point of view, linking the planning of the public space in the city-sea interface to different functional fields, in which the socio-recreational component plays a relevant role, generating positive chain effects on the different segments of urban coastal society (Sandifer et al. 2021).

9.2 Quantitative Approach for Studying the Urban Amphibious

Urban coastal society has high expectations of planning intervention in coastal public spaces aimed at expanding the possibilities of sociality and recreation in large maritime cities, as set out in the reflections that defined the research question. In this vein, the investigated methodological approach tends towards the development of psycho-social and perceptive reading techniques. For this type of analysis, it is needed to delve into social listening methodologies rooted in sociological theory. The aim is to understand the opinions of citizens about the issues of accessibility and fruition of the public spaces along the urban coastline, through an in-depth study of the typical socio-recreational uses of the average user, while at the same time interpreting the possible suggestions regarding the possible or desired transformations of the urban amphibious, in both a planning and environmental perspective.

Both the qualitative and quantitative approaches aim to gather information. However. while the former method relies mainly on observation to directly study the behaviour and manifest characteristics of a place or community, the latter allows for a better understanding of the motivations for certain habits and to explore the expectations and perceptions of the individuals involved regarding particular conditions (Cardano 2003). This is related to the lines of research defined above. In order to generalise the ideas of a section of the population that frequents urban shores, it may be appropriate to select a type of quantitative approach, suitable for processing a large number of data that are as objective as possible. Quantitative research relates the basic theory and its application in logical methodological procedures aimed at verifying the theoretical assumptions through the collection of empirical information from field surveys (Corbetta 2014). On this basis, it is possible to translate concepts into empirical variables, which can be measured and studied, even before disseminating the survey tool. In a similar way, for the purposes of the quantitative study, the relationship between users and the reference context is preliminary taken into consideration for its influence on the reactions of the analysed individuals, since it can be overlooked or manipulated in a controlled manner within a certain limit; it follows that even the physical and psychological interaction between the scholar and the interviewed users should be scientifically neutral and as limited as possible (Cardano and Ortalda 2016).

Through the sample survey, the quantitative method thus allows to gather information by interviewing users of coastal society, who represent a peculiar segment of the urban community, through a standardised procedure that aims to highlight the relationships between the hypothesised variables. One of the most popular quantitative survey method in sociological studies is the questionnaire, which has been selected as the survey instrument for this aspect of the work (Beatty et al. 2019). The first phase of building the survey instrument involved an in-depth theoretical investigation from which the abovementioned research questions emerged. If the questionnaire is structured in an efficient and detailed manner, it can in fact serve as an efficient information base for technicians dedicated to spatial development and as a guide for policy makers that could benefit of such data for taking informed and targeted decisions: this kind of methodology allows for the collection of information in a standardised way on the basis of well-defined demographic samples from which results can be deduced and related to broader social groups. The theoretical research has thus organically connected a coherent set of hypothesis regarding the underlying study questions, with a certain level of abstraction that can be generalised with respect to the wider context of the city-sea interface, in order to draw valid and logical reasonings (Bailey 2006). Indeed, if the base theory may assume a broader point of view on a certain topic, the related operational hypotheses must be specific and consistent with the selected case study. This is followed by the design of the methodological structure of the psycho-social research, with the definition of the questionnaire itself: its dissemination is necessary to collect data that will then be organised and analysed with the target of providing an inductive interpretation of the obtained results (Saris and Gallhofer 2007).

With regard to the formulation of the possible answers, it has been chosen an additive scale, specifically a *Likert scale*. This is a measurement scale that assesses the level of agreement or disagreement of the user about a statement: the opinion about a fact can range from being completely in agreement on one side to being completely in disagreement on the other side; in the middle there is indifference or a lack of opinion on the subject. One of the highlights of the Likert scale is that it succeeds in capturing the nuances of the ideas of users, since it usually consists of three, five or seven points (Memoli 2004). This instrument has probably become the most popular format among scaling methods aimed at quantifying public opinion on any given issue: it is possible to measure how much someone agrees with an assertion, as mentioned above, but also to quantify the frequency with regard to certain activities, establishing a time range from 'often' to 'never' (Nemoto and Beglar 2014). However, it should be noted that this method, while having the ability to measure the degree of agreement with an opinion or the frequency of a certain action, can be negatively influenced by the tendency of the interviewees to agree with any proposed statement, without giving too much weight to the answers they give but trying to appear in line with the interviewer's thinking: it is therefore necessary to present the questions in an aseptic and balanced manner, in order to avoid to express personal opinions or feelings within the data collection process[1] (Robinson 2014). Keeping these concepts in mind, for the purposes of the present book, the usefulness of this method is therefore the possibility of achieving simple results, as numbers and graphs, which enables to generate a hierarchy of values for the individual questions, coherently with the simple structure of the questionnaire model elaborated in this work. In the hypothesised questionnaire frame, five-point scales are proposed, in order to have the following possibilities: (1) *Not at all*—(2) *A little*—(3) *Indifferent*—(4) *Quite much*—(5) *Very much*. This allows for a more specific

[1] In order to minimise misunderstandings, it is necessary to avoid adversative conjunctions that presuppose the presence of two ideas within the same question: these are referred to as *double-barrelled questions* (Wolfe and Smith 2007).

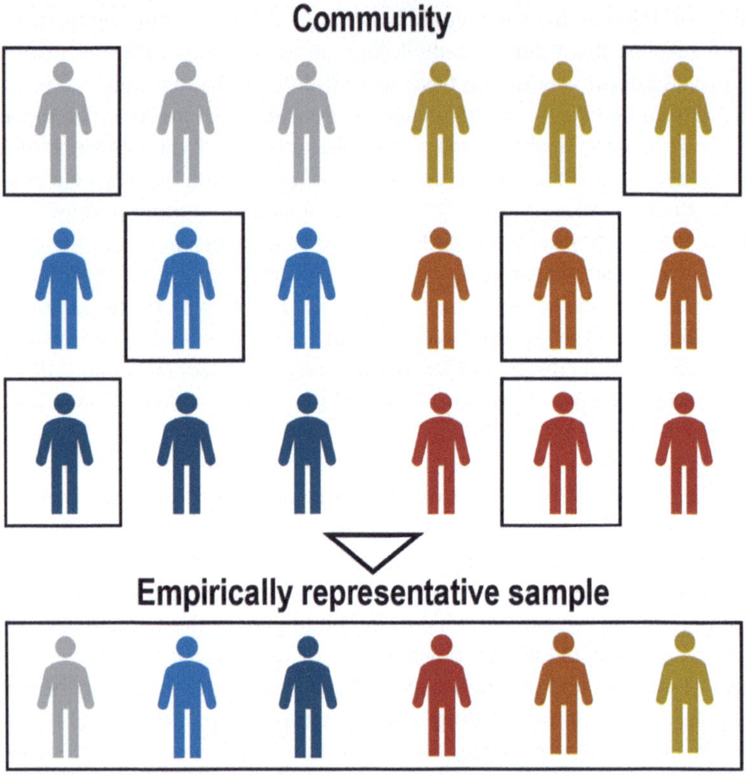

Fig. 9.1 The concept of representativeness is expressed through the selection of users within the sample in the widest possible and varied way: it empirically reflects habits and perceptions of a representative portion of coastal society (elaboration of the author)

expression of opinions by the users; at the same time, it is easy to organise the gathered data for the subsequent processing phase (Joshi et al. 2015).

In order to better understand the modalities and needs of the city-sea interface, with regard to social demand and perception, it emerges the methodological critical issue of precisely identifying coastal users, also on the basis of their place of provenance: in fact, it is not easy to assess whether all users of the coastal society come from neighbourhoods along the coast, from the reference city or even from other cities. The sampling error depends on the size of the analysed sample: if it is too small, there is the risk that the analysis could be biased. However, it is not only the size of the sample that determines its validity, and this parameter can still be in conflict with the accuracy of the sampling procedure (Corbetta 2014). It is therefore necessary to seek a balance between the two elements, bearing in mind that accuracy is the main quality for quantitative research: it is possible, in fact, to reach an extremely large number of users to interview, yet it will be almost impossible to identify every single component of the reference community, in this case the urban coastal society; on the contrary, it is plausible to identify and thoroughly analyse a small but nevertheless well varied sample that will be sufficiently wide to provide appreciable data (Sharma 2017).

It is now introduced a peculiarity of the questionnaire and the quantitative approach, namely the concept of representativeness, in relation to the sampling of studied subjects. A sample can be defined as representative if it provides a small but unbiased picture of the population (Grafström and Schelin 2013) (Fig. 9.1). It is thus stated that the distribution of a certain number of variables within a community does not have to be necessarily related to the whole population: it is sufficient to select a limited sample, based on certain criteria, in order to obtain empirically valid and generalisable results for the city-sea interface. It is clear that the representativeness of the sample will not be absolute but will depend on the variables initially chosen: the sample will be representative with respect to the selected parameters in relation to a given social phenomenon, admitting the existence of other variables. However, since it is impossible to codify all of them and apply them to the entire demographic spectrum, applying the survey methodology in this way will still produce valid results: although it will not have a statistical correspondence with the demographic composition of the entire coastal society, the reliability of the research method will be achieved anyway, allowing to generalise the results obtained from the questionnaires (De Singly 2020). On the basis of this reasoning, it is therefore proposed to analyse a reduced sample: its composition is defined according to gender and age groups (under 20 years, between 20 and 40 years, between 40 and 60 years, over 60 years); in addition to this, there are additional characteristics such as level of education, current employment, place of residence and time period in which users have lived in a certain place (Moro 2011).

9.3 Structure of the Survey Instrument: The Questionnaire

With regard to the hypotheses underlying the analysis, it is stated that, although it is clear that cultural level and personal wealth influence the way people experience the coast, it is less obvious whether these factors bind users to particular places on the coastal interface, for example, in relation to attachment to place. In this sense, it is possible to hypothesise that the distance of residence from certain points of the urban sea may exert a different influence on members of urban coastal society: plausibly, places that are not well-known or attractive from a tourist point of view might be frequented and appreciated mainly by users living nearby, regardless of social classes (Pasanen et al. 2019). Therefore, a structured questionnaire is proposed, divided into three main sections (*A. Accessibility of urban blue spaces, B. Land and water uses in the city-sea interface, C. Perception and future scenarios of the urban coast*), with a total of 12 questions to be submitted to the local users.

Section "*A. Accessibility of urban blue spaces*" aims to analyse the theme of the city-sea interface, according to the importance that users attribute to direct contact with the maritime resource, defining a scale of frequency in which citizens visit urban shores, in order to establish whether the latter can really be considered a gateway to the city sea. The proposed questions are the following ones:

1. In the city, how important is it for you to have physical contact with water (*e.g. swimming, bathing, fishing, diving*)?
2. In the city, how often do you go to the coast for recreation?
3. From where you live, how easy is it to reach the shore in the city?
4. In the city, what is your preferred means of transport to the reach the shore?
5. In the city, do you think that access to the coast is limited by physical barriers (gates, barbed wire, fences), pollution or poor water quality, legal constraints (e.g. regulations, forbidden areas)?

The influence that light or heavy transport systems have on the relationship that users establish with the coast is assessed, but also whether the presence of physical or legal barriers can actually prevent citizens from reaching these very important recreational spaces; moreover, it is studied if the problem of environmental pollution destabilises the social uses of the urban sea (Biernacka and Kronenberg 2019). As described above, the issue of accessibility and enjoyment of the city-sea interface, in relation to the multiplicity of its uses, is closely connected to the social criticalities that affect contemporary urban coasts: it is important to rethink the functions of these places in relation to the most relevant social components, in order to favour their use by citizens (Carta and Ronsivalle 2016). Land connections, the presence of soft mobility systems and eventual accesses to the sea influence the activities and fruition of urban shores: studying how city flows converge towards the urban coasts can help planners and policy makers to interpret the desires of users in relation to local public spaces, within which the littoral community can actually satisfy its social and recreational needs through the services provided by the city-sea interface. Indeed, citizens benefit from the proximity to green and blue areas that allow for daily social activities, such as bathing or walking along the coast: it would be useful to examine the possibility of access to such coastal public spaces that allow the urban coastal society to get in touch with the city's maritime sphere (Cortinovis et al. 2018). For this reason, it is interesting to understand the extent to which users frequent the coastal areas of the city-sea interface: it seems relevant to study whether the presence of physical or legal obstacles may actually prevent them from reaching these very important recreational spaces and whether the issue of environmental pollution may constitute a destabilising element in the social uses of the coast and the urban sea. In this sense, analysing how coastal benefits relate to social demand may support a more inclusive and equal urban planning, aimed at a better distribution of services along the urban amphibious (Geneletti et al. 2020).

The functional competition existing between the uses related to the socio-recreational demand of the coastal community and the development of residential, economic and logistic functions along the city-sea interface addresses the questions of section "*B. Land and sea uses in the city-sea interface*" in order to understand whether the urban coastline offers a sufficient amount of public natural and open spaces dedicated to sociality and an adequate ecological quality of the coastal green–blue infrastructure: for this reason, it is interesting to analyse whether citizens consider these areas to be effectively well equipped and how they would implement their effectiveness in the short term, stimulating their reasonings regarding the fruition of the urban littoral but also of the sea in their city (Badami and Ronsivalle 2008). In particular, the following questions have been posed to the users:

6. Here in the city, are there enough public areas, in your opinion?
7. Here in the city, what facilities would you like to develop or improve along the seashore in the future?
8. Today, to your knowledge, what social or recreational activities exist or take place here along the seashore?
9. In the future, what activities would you like to see promoted or allowed here along the seashore?

The second section of the questionnaire therefore aims to study the functional conformation of the public areas located along the urban coastline, in relation to the granted services and social uses: it aims to stimulate the citizens' imagination on the theme of the current use of the urban amphibious and how it could evolve from a functional point of view, evaluating the existing and desired offer of socio-recreational activities that the city-sea interface actually provides for the urban coastal society. There is in fact a considerable influence on the activities and needs of users that must be oriented towards quality of life and reduction of social stress, especially at the local scale within proximity public space. Mainly at the micro-urban level, there is a clear need for open areas and natural and semi-natural places that allow the users to get in touch with the coastal-maritime environment, despite critical building and spatial issues, urban barriers, increasing contemporary population density and environmental problems. The land and water spaces of the interface offer the possibility of functionally regenerate the social and recreational network of the urban coastline, providing theoretical and operational insights for a real functional sustainability at urban and neighbourhood scales, guaranteeing important transformative opportunities in favour of the urban coastal society (Tan and Jim 2017): the questions therefore aim to translate the demand for recreational uses and the need for contact with the coast into efficient services and planning indications, following the suggestions that users can provide if properly guided and addressed.

Finally, if the public space offers a privileged meeting point within which the various social groups can interact, especially along urban coastal areas, historically a place of great sociability, it is also true that the condition of the urban coastline can influence the way it is perceived by the user, in relation to the fragmented and uneven conformation of the current urban amphibious (Nicholls et al. 2015): because of this, the questionnaire section "*C. Perception and future scenarios of the urban coast*" investigates how much the landscape and environmental aspects of the shore (considering both its terrestrial and maritime spheres) influence the community's common understanding of littoral issues and evolution. The questions proposed in this section of the questionnaire are listed below:

10. Here, what is the landscape value of the coast for the built-urbanized environment and for the natural environment, according to you?

11. In your opinion, how could this area evolve in five years in an environmental perspective?
12. How would you like this coastal area to evolve over the next five years to increase socio-recreational possibilities?

As described above, coastal urban blue spaces are places that provide direct and indirect access to urban water, similar to therapeutic landscapes, because they improve social cohesion, offering psychophysical benefits to the coastal community, which tends to prefer littoral areas to other urban green spaces (Bell et al. 2015). It is therefore interesting to understand citizens' attitudes towards these peculiar public areas, in relation to the possible recreational activities, the landscape value and the sensory stimulation by the coastal environment: the perceptive effect of maritime amenities, channelled by urban shorelines, brings positive feelings and a general well-being, creating interest and attachment to place in urban coastal society (Hawthorne et al. 2022). It is therefore necessary to define which are the most relevant factors to influence the development of the city-sea interface, with a focus on the implications for well-being and sociality: the coastal environment stimulates users, who in turn interact with each other and their surroundings through functions that need to be sustainably organised within urban coastal planning (Mishra et al. 2020); littoral design must also take into account quantitative and qualitative parameters to improve coastal public spaces, according to flexible frameworks to face contemporary environmental issues as well as social demand for urban services. In addition, the evolution of the city shore is a particularly relevant issue, within which the perspective of its direct user is essential.

From a spatial and functional point of view, the connection with the concept of the urban amphibious is evident, because of its nature of contact area between land and sea of the city that is constantly reshaped and reorganised according to the two systems. Its adaptability is linked to the possibility of strengthening the existing socio-recreational functions, but also of establishing new ones, in closer relationship with the maritime element. In this sense, the questions in the section aim to highlight the transformation that users consider most plausible and desirable, analysing their knowledge and involvement in the social issues but also in the economic and environmental constraints that characterise the city-sea interface (Fig. 9.2).

UNIVERSITÉ FRANCO ITALIENNE — UNIVERSITÀ ITALO FRANCESE — CAMPUS FRANCE — UNIVERSITÀ DEGLI STUDI DI TORINO

International research programme
"GALILEO 2021"

Code:______________

RE-SEA-OURCING CITY
City-sea interface as a resource for people: urban regeneration in the context of ecological transition

QUESTIONNAIRE
Accessibility, usability and future scenarios of the urban coasts

Date: ______________
Hour: ______________
Place: ______________

Consiglio Nazionale delle Ricerche — IRISS — DIARC — Aix-Marseille Université — CNRS

1. In the city, how much is important for you to have physical contact with the sea (*e.g. swimming, bathing, fishing, snorkelling*)?

1.Not at all	2.A little	3.Indifferent	4.Quite much	5.Very much

2. In the city, how often do you go to the coast for recreation?

- Never or rarely (*less than twice a month*)
- Sometimes (*about once a week*)
- Often (*about twice a week*)
- Very often (*at least three times a week*)

3. From where you live, how easy is it to reach the shore in the city?

1.Very hard / 2.Hard / 3.Within the norm / 4.Easy / 5.Very easy	
By walk	
Bicycle	
Bus	
Electric tram	
Metro	
Car	
Scooter / motorbike	

4. In the city, what is the means of transport to reach the shore that you prefer? *Select one option from the previous ones*

____

5. In the city, do you think shore access is limited by:

1.Not at all / 2.A little / 3.Indifferent / 4.Quite much / 5.Very much	
Physical barriers (*e.g. gates, fences, barbed wire*)	
Pollution / bad quality of water	
Legal constraints (*e.g. regulations, forbidden areas*)	

6. Here in the city, are there enough of the following public areas along the coast, according to you?

1.Insufficient / 2.A few / 3.Enough / 4.A lot / 5.More than enough	
Promenade / pedestrian areas	
Parks (*e.g. natural area along the shore, gardens*)	
Squares (*e.g. public plazas*)	
Sport areas (*e.g. playgrounds, outdoor training areas*)	
Jettys / water walkpath (*e.g. wood platform*)	
Beaches	

7. Here in the city, what facilities would you like to be developed or improved along the seashore in the future?

1.Not at all / 2.A little / 3.Indifferent / 4.Quite much / 5.Very much	
Restaurants, bars	
Shops	
Rental of nautical equipment (*e.g. kayaks*)	
Leisure docks	
Outdoor swimming pool	
Museums / cultural centres	
Submarine museums	
Improvement of water quality	
Improvement of marine biodiversity	
More vegetated areas on the coast (e.g. gardens, parks)	

8. Today, to your knowledge, what social or recreational activity exists or is taking place here along the seashore?

Strolling	Yes	No
Running / Jogging	Yes	No
Cycling	Yes	No
Visiting street markets	Yes	No
Sport activity (*e.g. outdoor gymnastics*)	Yes	No
Swimming / bathing	Yes	No
Rowing / sailing boats	Yes	No
Diving / Snorkelling	Yes	No
Fishing	Yes	No
Guided tours	Yes	No
Cultural or political events	Yes	No

9. In the future, what activities would you like to be promoted or allowed here along the seashore?

1.Not at all / 2.A little / 3.Indifferent / 4.Quite much / 5.Very much	
Strolling	
Running / Jogging	
Cycling	
Visiting street markets	
Sport activity (*e.g. outdoor gymnastics*)	
Swimming / bathing	
Rowing / sailing boats	
Diving / Snorkelling	
Fishing	
Guided tours	
Cultural or political events	

10. Here, what is the landscape value of the coast, according to you?

1.Very low / 2.Low / 3.Neutral / 4.High / 5.Very high	
Built-urbanized environment	
Natural environment	

11. In your opinion, how this area is likely to evolve in the next 5 years from an environmental perspective?

1.Not likely / 2.Hardly likely / 3.Likely / 4.Quite likely / 5.Very likely	
Sea-level rise	
Strong coastal erosion	
Frequent coastal floods	

12. How would you like this coastal area to evolve over the next 5 years to improve socio-recreational possibilities?

1.Not at all / 2.A little / 3.Indifferent / 4.Quite much / 5.Very much	
New land reclaimed from the sea	
Floating structures (*e.g. platforms, floating solariums*)	
New leisure docks	
New public leisure areas (*e.g. squares, outdoor sport areas*)	
More commercial-oriented areas (*e.g. shops, restaurants*)	
New areas or facilities for bathing	

Personal informations

Year of birth:______________
Place of birth:______________
Gender:______________
City of living:______________
Neighbourhood of living:______________
Since when do you live there?______________
Educational level:______________
Current job:______________

Additional notes

Fig. 9.2 Structure of the questionnaire for Marseille and Naples' city-sea interfaces (elaboration of the author in collaboration with CNRS ESPACE—UMR 7300)

References

Badami A, Ronsivalle D (eds) (2008) Città d'acqua. Risorse culturali e sviluppo urbano nei waterfront. Aracne Editrice, Rome

Bailey K (2006) Metodi della ricerca sociale. I principi fondamentali. Il Mulino, Bologna

Beatty P, Collins D, Kaye L, Padilla L, Willis G, Wilmot A (eds) (2019) Advances in Questionnaire Design, Development, Evaluation and Testing. Wiley, Hoboken

Bell SL, Phoenix C, Lovell R, Wheeler BW (2015) Seeking everyday wellbeing: The coast as a therapeutic landscape. Social Science & Medicine 142: 56–67

Biernacka M, Kronenberg J (2019) Urban Green Space Availability, Accessibility and Attractiveness, and the Delivery of Ecosystem Services. Cities and the Environment (CATE) 12(1)

Bratman GN, Olvera-Alvarez HA, Gross JJ (2021) The affective benefits of nature exposure. Social and Personality Psychology Compass 15(8): 1–22

Cassatella C (2011) Assessing Visual and Social Perceptions of Landscape. In Cassatella C, Peano A (eds) Landscape Indicators. Springer, Dordrecht, p. 105–140

Cardano M (2003) Tecniche di ricerca qualitativa. Percorsi di ricerca nelle scienze sociali. Carocci Editore, Rome

Cardano M, Ortalda F (2016) Metodologia della ricerca psicosociale. UTET Università, Turin

Carta M, Ronsivalle D (eds) (2016) The Fluid City Paradigm. Waterfront regeneration as an Urban Renewal Strategy. Springer, London

Collado S, Staats H, Corraliza JA, Hartig T (2017) Restorative Environments and Health. In Fleury-Bahi G, Pol E, Navarro O (eds) Handbook of Environmental Psychology and Quality of Life Research. Springer, Berlin, p. 127–148

Corbetta P (2014) Metodologia e tecniche della ricerca sociale. Il Mulino, Bologna

Cortinovis C, Zulian G, Geneletti D (2018) Assessing nature-based recreation to support urban green infrastructure planning in Trento (Italy). Land 7(112)

De Singly F (2020) Le questionnaire. Armand Colin, Malakoff

Geneletti D, Cortinovis C, Zardo L, Esmail BA (2020), Planning for Eco-system Services in Cities. Springer, Cham

Georgiou M, Morison G, Smith N, Tieges Z, Chastin S (2021) Mechanisms of impact of blue spaces on human health: A systematic literature review and meta-analysis. International Journal of Environmental Research and Public Health 18(5): 1–41

Grafström A, Schelin L (2013) How to Select Representative Samples. Scandinavian Journal of Statistics 41(2): 277–290

Hawthorne TL, Toohy KR, Yang B, Graham L, Lorenzo EM, Torres H, McDonald M, Rivera F, Bouck K, Walters LJ (2022) Mapping Emotional Attachment as a Measure of Sense of Place to Identify Coastal Restoration Priority Areas. Applied Geography 138

Hipp JA, Ogunseitan OA (2011) Effect of environmental conditions on perceived psychological restorativeness of coastal parks. Journal of Environmental Psychology 31(4): 421–429

Hooyberg A, Michels N, Allaert J, Vandegehuchte MV, Everaert G, De Henauw S, Roose H (2022) 'Blue' coasts: Unravelling the perceived restorativeness of coastal environments and the influence of their components. Landscape and Urban Planning 228

Joshi A, Kale S, Chandel S, Pal DK (2015) Likert Scale. Explored and Explained. British Journal of Applied Science & Technology 7(4): 396–403

Luginbühl Y (2017) Landscape Iconography and Perception. In Richardson D, Castree N, Goodchild MF, Kobayashi A, Liu W, Marston RA (eds) International Encyclopedia of Geography: People, the Earth, Environment and Technology. Wiley, Hoboken, p.1–8

Memoli R (2004) Strategie e strumenti della ricerca sociale. FrancoAngeli, Milan

Mishra HS, Bell S, Vassiljev P, Kuhlmann F, Niin G, Grelli J (2020) The development of a tool for assessing the environmental qualities of urban blue spaces. Urban Forestry & Urban Greening 49

Moreno C (2020) Vie urbaine et proximité à l'heure du COVID-19. Editions de l'Observatoire, Paris

Moro G (2011) La valutazione possibile. Metodi e casi. Carocci Editore, Rome

Nemoto T, Beglar D (2014) Developing Likert-scale questionnaires. In Sonda N, Krause A (eds) JALT2013 Conference Proceedings. JALT, Tokyo

Nicholls RJ, Stive MJF, Tol RSJ (2015) Coping with Coastal Change. In Masselink G, Gehrels R (eds) Coastal Environments and Global Change. Wiley, Hoboken

Pasanen TP, White MP, Wheeler BW, Garrett JK, Elliott LR (2019) Neighbourhood blue space, health and wellbeing: The mediating role of different types of physical activity. Environmetal International 131

Robinson J (2014) Likert Scale. In Michalos AC (ed) Encyclopedia of Quality of Life and Well-Being Research. Springer, Berlin, p. 3620–3621

Sandifer PA, Keener P, Scott GI, Porter DE (2021) Oceans and Human Health and the New Blue Economy. In Hotaling L, Spinrad RW (eds) Preparing a Workforce for the New Blue Economy. Elsevier, Amsterdam, p. 213–236

Saris WE, Gallhofer IN (eds) (2007) Design, Evaluation, and Analysis of Questionnaires for Survey Research. Wiley, Hoboken

Severin MI, Vandegehuchte MB, Hooyberg A, Buysse A, Raes F, Everaert G (2021), Influence of the Belgian Coast on Well-Being During the COVID-19 Pandemic. Psychologica Belgica 61(1): 284–295

Sharma G (2017) Pros and cons of different sampling techniques. International Journal of Applied Research 3(7): 749–752

Subiza-Pérez M, Vozmediano L, San Juan C (2020) Green and blue settings as providers of mental health ecosystem services: Comparing urban beaches and parks and building a predictive model of psychological restoration. Landscape and Urban Planning 204

Tan PY, Jim CY (eds) (2017) Greening Cities. Forms and functions. Springer, Berlin

Wolfe EW, Smith Jr EV (2007) Instrument development tools and activities for measure validation using Rasch models: Part I - Instrument development tools. Journal of Applied Measurement 8: 97–123

Chapter 10
Methodological Framework for Socio-Environmental Benchmarking of City-Sea Interfaces

Abstract The considerations made so far have contributed to the definition of a methodological framework dedicated to the study and understanding of the city-sea interface and its socio-ecological features. In particular, the validity of this method in benchmarking processes between different coastal cities is emphasised. The relevance of this approach lies in the definition of shared knowledge concerning the perceptions of the urban coastal society and the dynamics inherent in the governance of the maritime territory. Indeed, the comparison between coasts that are similar in terms of functionality and socio-ecological spheres can serve as an efficient reading key to common problems.

Keywords Methodological framework · Benchmarking · Socio-ecological features · Coastal areas · Complexity · Maritime governance

10.1 Understanding the Complexity of Urban Coastal Space Through Comparison

As previously discussed, the coastal space of the contemporary city appears to be a complex field within which numerous features can be studied with a dedicated approach: in fact, the spatial and functional fragmentation and dispersion recorded in such areas is evident, especially over the last few decades; the structure of compact cities has been gradually dilated into widespread settlements due to the progressive growth of modern urban fabric and increasingly articulated and mystifying functional dynamics. The demographic increase is linked to social needs that are not always satisfied by appropriate public spaces and community services, as the city and its shore are often shaped by economic stakes aimed at profit rather than the well-being of users. Added to this is the issue of climate change: environmental adaptation requires an assessment of future scenarios of urban evolution and the development of strategies and plans to allow human coexistence with ongoing ecological transformations. Therefore, proactive urban planning is necessary in order to face environmental risks and develop preventive actions and interventions (Cooper and Lemckert 2012).

I. Pistone, *Urban Coasts in Socio-ecological Transition*,
https://doi.org/10.1007/978-3-031-70783-4_10

In the Euro-Mediterranean context, it has also been pointed out that urban planning within the city-sea interface tends to be uneven, both from the point of view of socio-recreational uses and environmental protection. It is evident that the proximity of settlements to the sea and the availability of areas suitable for bathing represent strong tourist attractors, modelling the evolution of the urban amphibious which tends to transform and adapt to contingent and spatial needs, due to its dynamic nature. The increase of the social demand for coastal space correlated to the change of the littoral territory represents one of the main problems linking planning activities and environmental sustainability for maritime areas, while factors such as urbanisation, economics, transport, tourism and recreational use of the interface can alter the balance of land-sea interactions (Colannino and Roca Cladera 2008).

It seems appropriate to enrich the research process through a methodological framework that considers the different actors of urban coastal society, their interrelationships, needs and perceptions as relevant inputs to comparatively assess urban-coastal dynamics. Benchmarking is a process that allows entities, institutions, spatial and functional organisations to improve themselves through the comparison of existing characters, ideas and solutions: in short, it basically allows them to learn from each other from similar situations, using the experience gained in other places. The purpose of the comparison is to identify ideas and directions for improvements, on the basis of managerial and planning adaptation to existing situations in order to implement the public space along the city-sea interface: in this sense, it can represent an added value for coastal strategies, highlighting criticalities and processes that can be revised or reconfirmed. The analysis of data found in different points of the Euro-Mediterranean basin can therefore open up scenarios of technical and scientific cooperation and collaboration, linking critical knowledge from various fields (Lanzetta 2020).

The study of existing governance processes and regulations intends to provide a picture of the evolution of policies and decision-making tools within the coastal cities analysed, in order to understand how littoral interface planning is carried out and what lines of implementation and social integration are envisaged: this can provide important information regarding the institutional relationships between the actors involved in urban littoral development, highlighting possible similarities and discrepancies in the case studies, assessing the degree of participation of urban coastal society and the actual interest of the municipality in environmental and socio-recreational but also economic and logistical terms.

The relationship between the sea, the coast and the existing spaces and infrastructures, whether civil, port or public, must be analysed, taking into consideration not only the most relevant neighbourhoods along the urban coast and the areas adjacent to harbours, but also the peripheral sites far from the city centre, areas with an ancient productive vocation now in a state of decay, and spaces of the city-sea interface where strong community and environmental stakes insist. Finally, benchmarking must consider the parameter of social expectation and perception, interpreting the different needs of the coastal community and pointing out how the desire for landscape visibility and functional accessibility to the sea resource affects the way different city-sea interfaces are understood, within the same Euro-Mediterranean basin (Cattedra 2011). In addition to this, the method presented can be inserted within a broader process of urban planning and design of the city-sea interface, in relation to the social demand for coastal public space: the multidisciplinary context of the present work has allowed to elaborate a detailed approach, which becomes an important basis for future planning and management development by urban planners and local policy makers, from an integrated perspective.

10.2 Methodological Notes

In order to provide adequate guidance for possible comparative operations between different city-sea interfaces, it is necessary to clarify some principles underlying the application of the tools and approaches illustrated in the previous chapters.

Once the urban form of the chosen cities has been studied, it is possible to proceed with a *normative analysis* in order to study the existing legal framework that characterises them. The direction of the approach will clearly follow the indications of the research question, regarding the socio-ecological planning of coastal public space within the city-sea interface. The methodological criteria were chosen according to the aspects considered most suitable for the legal investigation. The normative analysis therefore envisages the collection and study of existing legislative frameworks in the urban coasts to compare, from a social, cultural, environmental and management point of view: in particular, the analysis of the city-sea interface must focus on directives relating to environmental and landscape value, water quality in terms of bathing, but also the risks of erosion, flooding and other critical aspects of climate change; it is also important to search for regulations and urban planning tools that regulate accessibility to the urban coast, in relation to socio-recreational practices, fruition and social safety of places. Furthermore, the documents analysed should focus mainly on the local scale, in order to verify the presence of an adequate governance of the city-sea interface, from the point of view of existing instruments and from a participatory and community perspective.

Concerning the *geo-spatial approach*, the realisation of the spatial data model must consider the spatial accuracy of the collected information. Within the proposed framework, most of the data concern land uses that are difficult to find from existing databases, or in any case available only at a very large scale: for this reason, digitisation is necessary for redrawing elements by means of source images in raster format. These files must therefore possess the highest possible degree of resolution to guarantee efficient decoding of the coastal interface in the various case studies to be compared. For this, it is appropriate to introduce the concept of the minimum mapping unit (MMU), which is the specific size of the smallest element that is reliably identified within a map (Buckley 2008). This is a very important parameter in both data collection and actual map production. Indeed, it allows a skimming process among the data to be analysed and produced, also in relation to the scale of the final output (Peterson 2021). Depending on the scale considered, spatial data may offer different information on the characteristics studied and the interrelationships that are established: consequently, since scale influences any study of geographic data, spatial analysis needs to evaluate the scale components through minimum mapping units (Álvarez 2019). In the specific case of the spatial data model proposed in this book, the reference base will generally consist of aerial photographs of the study area: the main research object concerns the socio-recreational use of coastal resources in the city-sea interface, consequently the aim is to investigate the reduced space in which the urban coastal society acts and comes into contact with the urban sea. For this reason, it is desirable to refer to very low values of MMUs: the small public spaces and the reduced natural area along the urban coast constitute the founding elements of the urban green–blue infrastructure at the basis of the socio-ecological planning of the city-sea interface.

The *psycho-social approach* introduced here is realised through the definition of a questionnaire, as described above. The dissemination of this survey instrument takes place in the form of a structured interview, working directly on the field and interacting with the users within the reference social sample. The individual interview thus allows the creation of a dialogical relationship between the researcher and the interviewee: this has several positive aspects because it gives the possibility to explore in depth the questionnaire's themes together with the user, dispelling doubts relating to a misinterpretation of the questions and possible answers (Corrao 2005). In fact, the cultural gap between expert and non-expert users is a common problem: although the structure of the questionnaire does not foresee correct or incorrect answers, it is still important that the requested information is adequately interpreted for the studying the social perception of the city-sea interface. In addition, an attempt is made to prevent possible biases of the studied social sample: since it is a numerically small social piece, it is indispensable that the information collected meets the socio-demographic criteria underlying the analysis. By dividing the reference urban coastline into spatial areas within which proportionally and equally collecting data with the fieldwork, it is possible to prevent this criticality. Massively disseminating the questionnaires through digital platforms could mean that the interviewee may be a different person than the one chosen in accordance with the social representativeness theory, with other unnecessary characteristics: in the case of a wide dissemination that is not precisely controlled, there is no way of knowing the exact extent of the error; furthermore, it may also happen that the questionnaire is filled in by a user included in the sociological survey spectrum, but with the help of a third party. The dissemination of the sample survey on the field can provide the necessary expert help to the users, without altering their answers.

According to the proposed analytical hypothesis, the consequentiality of these approaches can provide a support for the planning and management of the reference case studies by comparing the peculiar features of their city-sea interfaces, in relation to some specific areas in line with the research question. In particular, this book will present the benchmarking between two large Euro-Mediterranean cities, which will be discussed in the following chapters: Marseille and Naples.

References

Álvarez D (2019) Cartographic Scale and Minimum Mapping Unit Influence on LULC Modelling. In Ragia L, Laurini R, Rocha JG (eds) Proceedings of the 3rd International Conference on Geographical Information Systems Theory, Applications and Management (GISTAM 2017). Springer, Cham, p. 327–334

Buckley AR (2008) Minimum Mapping Unit (MMU). In Kemp KK (ed) Encyclopedia of Geographic Information Science. Sage, New York

Cattedra R (2011) Projet urbain et interface ville-port en Méditerranée. Perspectives pour une recherche comparative. Rives méditerranéennes 39: 81–102

Colannino N, Roca Cladera J (2008) Modelli di urbanizzazione costiera: morfologia e complessità strutturale, a scala urbana e territoriale, nella regione metropolitana di Barcellona. ACE – Arquitectura, Ciudad y Entorno 3(7): 273–294

Cooper JAG, Lemckert C (2012) Extreme sea-level rise and adaptation options for coastal resort cities: a qualitative assessment from the Gold Coast, Australia. Ocean & Coastal Management 64: 1–14

Corrao S (2005) L'intervista nella ricerca sociale. Quaderni di sociologia 38: 141–171

Lanzetta A (2020) Benchmarking: Understanding the Basics. In Toppetti F, Ferretti LV (eds) La cura delle città: politiche e progetti. Quodlibet, Macerata, p. 87–93

Peterson GN (2021) GIS Cartography. A Guide to Effective Map Design Routledge, London

Part III
Applying the Methodological Framework to the City-Sea Interface: The Cases of Marseille and Naples

Chapter 11
Marseille, Naples: European Urban Coasts in Comparison

Abstract A comparative analysis between Marseille and Naples is proposed. Both are strongly urbanised Euro-Mediterranean coastal areas with high socio-cultural and economic values. In the French case, the coast is characterised by strong tensions between logistic-port development and the intense connection between citizens and the urban sea, although projects and actions have been implemented to mend this fracture. In the Italian case, the criticalities are linked to weak coastal planning that clashes with community demands. This work therefore aims to understand whether the city-sea interfaces of Marseilles and Naples can really meet local socio-ecological demands.

Keywords Naples · Marseille · Urban coasts · Socio-ecological demands · Comparison · Euro-Mediterranean cities · Littoral public space

11.1 Motivation and Context of the Comparison

The scope of this work and the reflections so far expressed suggest the need to define specific case studies. In this sense, it is proposed a parallelism between two cities of great relevance in the social, economic and cultural context of the Mediterranean basin, namely Marseille and Naples, coastal urban settlements that show similarities from a functional, demographic and spatial point of view, even though they present significant differences.[1] The urban coast of Marseille is characterised by the conflict between port and productive areas, where social tensions amplify the interpretative complexity, and spaces that are more prone to satisfy the needs of users, in the southern part of the shore. Similarly, the urban coast of Naples is an emblematic case for the presence of criticalities linked to a weak littoral planning in a social perspective, even if the local coastal community is increasingly making its voice heard, especially in more peripheral areas of the Neapolitan shore. The aim

[1] The work can be framed within the international research project 'Re-SEA-ourcing CITY. City-sea interface as a resource for people: urban regeneration in the context of ecological transition', funded by the Université Franco-Italienne for the Galileo 2021 programme.

I. Pistone, *Urban Coasts in Socio-ecological Transition*,
https://doi.org/10.1007/978-3-031-70783-4_11

is therefore to understand whether the city-sea interfaces of Marseille and Naples are actually designed and managed to meet the socio-recreational expectations of citizens. For this reason, the work will delve into specific areas of the two cities, on the basis of similar characters that allow comparison. To answer this question, the research will analyse the urban dynamics and planning of these spaces, examining if public policies and documents guarantee the accessibility to coastal functions for all users and ensure permeability between the shore and the rest of the city, applying the abovementioned methodological framework.

Both are important Mediterranean cities with a similar size and an equally complex articulation of the city-sea interface: in fact, their urban shores present comparable zones, from a physical-morphological and functional point of view; in both cities, it can be noted a juxtaposition of industrial and port areas, urban areas with a naturalistic-residential value, urban beaches and dense urban areas in the heart of the city-sea interface. Even from a socioeconomic point of view, the two cities have similar difficulties, sharing the same problems of equal accessibility to the sea, although Naples has remained fundamentally static during the last fifty years, while Marseille has tried to respond to the demands of coastal society with targeted interventions and policies, proposing innovative solutions to regenerate its urban amphibious from a recreational, cultural and environmental point of view. For instance, the French city has developed the *Musée Subaquatique de Marseille,* in the *Anse des Catalans* area, the first underwater museum in Europe, and a system for the production of thalasso-thermal energy in the middle part of the coast, exploiting the natural warmth of the water for a clean and sustainable power source. However, both cities have considerable influence on their metropolitan area and can be considered as main Euro-Mediterranean poles (Bertoncello and Rodrigues-Malta 2003).

A spatial comparison between the two cities is then proposed, analysing the main features and functional conformation of their urban coastlines, as well as the regulatory-legal framework in Naples and Marseille; physical-morphological and functional data will be collected in order to apply the theory of the spatial data model, highlighting the main points for comparing the two coastal interfaces and their spaces of sociality; this will be followed by the application of the social analysis method, with the distribution of questionnaires in specific points along the Marseille and Naples coasts.

The objective is to describe the influence exerted by morphological aspects of the urban coastal structure, economic-social elements and the actual presence of services and public spaces on the use that citizens make of their own city-sea interface, identifying the relationship between these factors and the socio-environmental development of urban shores. Since the perception of space is conditioned by social fruition and habits, based on the presence of functional activities and anthropic and natural elements shaping the contemporary urban coastline, the benchmarking aims to define the social representation that the mind of the users attribute to the city-sea interface in relation to their needs, an expression of the functional expectations of the urban coastal community.

11.2 Conformation and Social Characteristics of the Urban Coast of Marseille

Marseille is the second largest city in France in terms of urban size and population density. These features make it the main coastal socio-economic pole of the country, placing it among the most influential urban entities in the Mediterranean sea. The city's coastline is articulated around a central core dedicated to port-logistics functions, mostly inaccessible for the population. Northward is located a residential-recreational area, while the southern area of the coast is directly related to the historic centre and the *Vieux Port*, the ancient access from the sea, now a residential and tourist area.

These coastal areas are characterised by a fragmented and varied demographic composition in social, cultural and economic terms (Bertrand 2012). Marseille is an interesting study case from the perspective of coastal regeneration, as it has experienced, in various stages, the different coastal transformation phases common to most large coastal cities over the last fifty years, with planning and operational innovation at various scale and in different fields. The transformative impulse started as early as the 1960s, when in 1963 the plan for the development of Marseille metropolitan area displaced the urban expansion in the peripheral municipality of Fos-sur-Mer, separating productive and commercial port activities in liminal settlements from tertiary and tourist uses in the main urban pole of the territorial shore: this addressed the problem of the city's role in a vast urban region, but at the same time freed the Marseille urban structure from a cumbersome element such as the main harbour infrastructure (Zalio 1996). The 1990s then marked an important moment for the Marseille coast, with the launch of the national interest operation *Euroméditerranée,* which envisaged the spatial and functional recomposition of the southern side of the urban coastline, for socioeconomic, cultural and environmental purposes, aiming to reconcile the port vocation, in order to strengthen the role of the city within the Mediterranean basin, with urban regeneration and community demands, despite the overall turn towards economic profit and globalisation (Bertoncello and Dubois 2010). From an environmental point of view, Marseille relates to the sea differently depending on the geographical area. In the southern part, the urban fabric fades into a progressive gradient of naturalness; the port-productive area in the north is functionally confronted with the sea mainly through an artificialised coastline; in the urban core, the shore denotes a greater complexity, with a radial green structure articulated from the centre towards the periphery with different densities (Consalès et al. 2012).

It is conventional to divide the Marseille coastal basin, or *Rade*, into two juxtaposed sections, the wealthier *Rade Sud*, where the majority of social spaces are located, and the *Rade Nord*, with a more productive and logistic character (Fig. 11.1). The Rade Sud extends from the *Parc National des Calanque*, established in 2012 and considered the first park in Europe to be both a land and sea protected area within an urban settlement, and the *Vieux Port*, in the heart of Marseille city-sea interface. It includes the long and wide *Corniche* promenade, through which it is possible to reach the main coastal public spaces such as the *Plage des Catalans*, the *Plage des Prophètes*, and the *Parc Balnéaire du Prado*: as mentioned above, the latter was

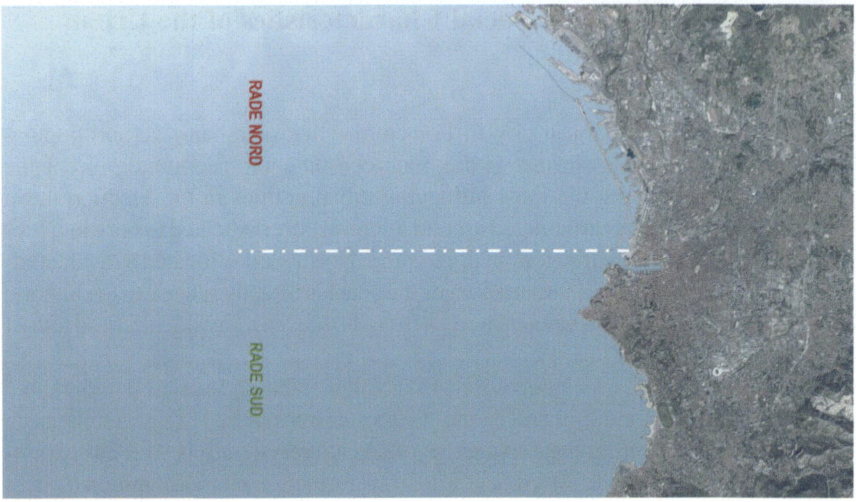

Fig. 11.1 Subdivision of the urban coast of Marseille into Rade Nord and Rade Sud (elaboration of the author)

realised in the 1960s and serves as a true urban terrace overlooking the sea, giving access to relevant urban beaches and extensive coastal green spaces. It is important to emphasise that these natural and semi-natural urban areas, harmoniously integrated into the urban fabric, have a public nature: it is evident the recreational-tourist and

Fig. 11.2 Users on the urban beach of Pointe Rouge, at the southern end of the Marseille city-sea interface; in the background the Parc du Prado (picture by the author)

Fig. 11.3 The long cycle-pedestrian promenade of the Corniche connects the main points of the Rade Sud of the city (picture by the author)

bathing character of the Rade Sud, along which it is possible to find high-end housing services and a high social status (Figs. 11.2 and 11.3).

In general, urban beaches and coastal open spaces are easily accessible and constitute an essential element of the cityscape of Marseille: anyway, even if this zone is frequented by users from different social backgrounds, the effects of its gradual regeneration, which began at the end of the last century, have favoured high-level commercial and restaurant activities, with consequent social fragmentation in the Rade Sud. The public authorities are dealing with this issue through targeted actions and regulations to allow social *mixité* and efficient coastal recreational management (Deboudt 2010) (Figs. 11.4 and 11.5).

The Rade Nord, on the other hand, is composed by the littoral areas between the historic fabric of the *Vieux Port*, where there are numerous cultural activities for the population as well as both planned and spontaneous leisure spaces (Fig. 11.4), and the *16th arrondissement,* the northernmost district of the city; in the middle, it is located the famous regeneration area of the *Euroméditerranée*. These areas demonstrate greater functional and social heterogeneity: it is easy to see that most of the available land and sea is occupied by functions related to the *Grand Port Maritime de Marseille* (GPMM). Over time, the coast has been completely redesigned by the port infrastructure, although several projects are trying to mend the relationship between harbour, shore and city (Fig. 11.5): within the *Euromediterranée* programme, it is possible to recall the abovementioned transformation of the former logistic J4 areas and buildings, near the neuralgic *Place de la Joliette*, with the realisation of important public works such as the MuCEM, a museum building connected to important elements of cultural heritage and vast public spaces along the coast, as well as *Les Terrasses du Port*, a shopping centre whose roof is conceived as a public square on

Fig. 11.4 The Vieux Port is one of the main social spaces of Marseille, into the historical core of the city. It is characterized by numerous leisure moorings (picture by the author)

Fig. 11.5 The J4 area hosts important cultural buildings (the CEREM and the MuCEM can be seen in the photo), as a result of the Euroméditerranée operation. Although the final result is a large coastal public space, there is still a very clear separation from the port, which occupies a wide part of the urban coast (picture by the author)

the sea in visual relation with the historical docks (Bertoncello and Rodrigues-Malta 2003). In any case, the neighbourhoods facing the harbour are strongly affected by inevitable functional constraints on the fruition of the urban sea, in relationship

Fig. 11.6 Cale de la Lave represents the only public access to the sea for boats in the whole city. As a result, it is often very crowded (picture by the author)

with this massive infrastructure, as well as by gentrification and the presence of economically weak social segments in degraded urban areas.

The case of the aforementioned *16th arrondissement* is emblematic: it is a former industrial area which, despite its recreational value, is extremely difficult to enjoy equally by locals: the Saumaty port and shipyard area and the numerous elite nautical clubs occupy almost the entire coastline, depriving the shore of almost any physical access to the sea and undermining socio-environmental justice, while the *Cale de la Lave* is the only public entrance for private boats not registered with clubs in the entire Marseille area (Fig. 11.6); long stretches of city-sea interface are designated for vehicular traffic only, lacking adequate space for a social fruition of the water resource; the *Plages de Corbière* complex is the only officially planned bathing spot in the area (Bertoncello and Hagel 2016) (Fig. 11.7). It is clear, therefore, that the feeling of social redemption in these places is as strong as the interest in urban transformation in the short term. The heterogeneity of the Marseille urban amphibious demonstrate the relevance of this case study but also how difficult is to adopt a univocal planning and management approach, in relation to the juxtaposition between port and productive areas, where social tensions amplify the complexity of functional interpretation, and spaces more inclined to host community functions, especially in the southern part of the littoral.

Fig. 11.7 The Base Nautique de Corbière, located in the homonymous area in the Rade Nord, represents an inclusive pecularity within Marseille as it is dedicated to the practice and teaching of water sports for users with disabilities (picture by the author)

11.3 Conformation and Social Characteristics of the Urban Coast of Naples

The coast of Naples, like all large and ancient European cities, is characterised by a millenary stratification of uses and transformations that results in a densely populated territory, with approximately one million inhabitants, with many natural and cultural resources but also degraded areas and competing uses. From the Pietrarsa museum on the eastern boundary with Portici to the Bagnoli area in the Gulf of Pozzuoli, the Neapolitan coastline is variously articulated and occupied for a large part by the port with its incessant activities, from the transport of goods to the mobilisation of passengers. According to ISTAT data, in 2020 the port of Naples has been the third in Italy for tourist transport[2]. The Port Authority of the Central Tyrrhenian Sea has studied passenger traffic in 2021 with approximately over 4,000,000 users, while container traffic has exceeded one million units: this testifies the relevance of the Neapolitan port, despite the negative effects of the pandemic period. Along the coastal interface, however, there is extremely poor accessibility to the urban sea for socio-recreational purposes, due to the low water quality caused by heavy logistical and productive activity, as well as to the privatisation of the coastline, which is detrimental to socio-environmental justice. Despite the progressive decommissioning of industrial and port facilities located along neuralgic areas of the urban coastline, there is still a lack of a virtuous interventions to positively exploit this planning opportunities: on the contrary, very often the extensive production of plans and projects has remained

[2] Data available at http://dati.istat.it/index.aspx?queryid.

on paper, without leaving tangible signs of coastal transformation except in minor moments in the recent history of the local city-sea interface. This leads to a marked social tension, exacerbated on the one hand by the presence of important stakeholders along the city shore, and on the other by society's desire to enjoy its urban sea on a neighbourhood scale (Papa 2010).

The progressive growth of the harbour infrastructure, started in the second half of the nineteenth century, has mainly occupied the eastern area (from the Beverello pier to San Giovanni a Teduccio), while the western part of the shore has been less influenced by this specific functional evolution over time, developing consistent values in terms of nature and landscape (Posillipo and Chiaia), with the exception of the Bagnoli plain, which since the early twentieth century has been occupied by the Ilva steel factory, definitively decommisioned in the 1990s. Similarly, also the eastern zone of San Giovanni a Teduccio has been for years site of industrial settlements that are now completely abandoned and only partially reconverted.

It is possible to ideally divide the Neapolitan coastline into various segments (Fig. 11.8). The eastern end is occupied by the area of San Giovanni a Teduccio, which for a long time hosted industrial plants and shows community conflicts from a socio-environmental point of view, in relation to the desire of local users to exploit more their littoral zone, even if the poor ecological condition of the water and the urban beach constitute an inevitable constraint (Palmentieri 2016). In sequence, there is the city port, which is composed by three subsections, namely

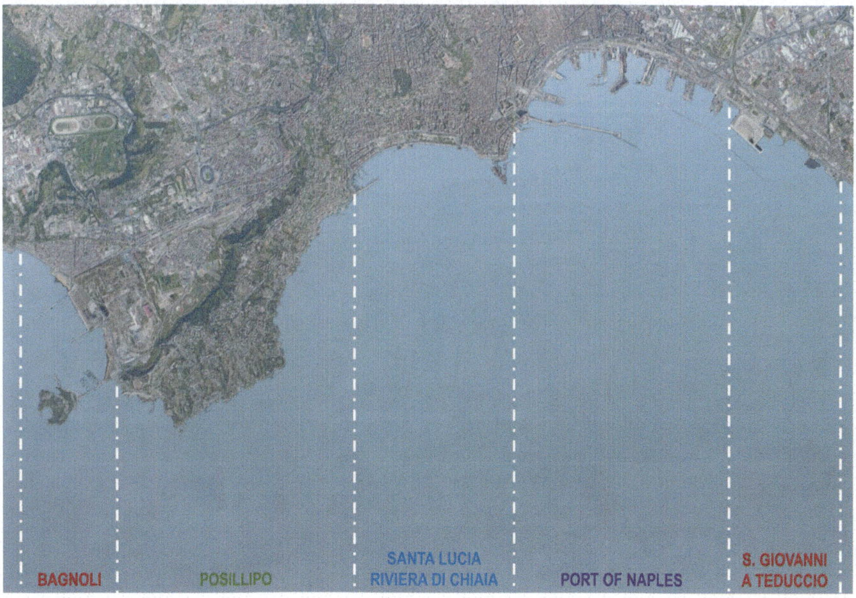

Fig. 11.8 Subdivision of the city-sea interface of Naples into spatial-functional areas (elaboration of the author)

the tourist port, that is currently under development, followed by the productive-commercial area and finally by the historical port of Naples, characterised by a large tourist flow: permeability towards the water mirror is clearly limited to a minimum due to the harbour infrastructure. The third section of the coastline includes the historic fabric of Santa Lucia and the Riviera di Chiaia, up to the Mergellina area: the wide stretch of coastline possess a predominantly social and residential character; this littoral area presents buildings with high architectural value, cultural features and historic gardens, such as the *Villa Comunale* and the Molosiglio gardens, which act as a hinge with the historic harbour and have a marked socio-recreational potential for the population, despite the conflicting interests with port and military institutions present nearby. The zone is not well equipped for bathing activities or direct enjoyment of the sea, although some parts of the water body would allow this type of function: this has led the population to exploit the existing localised beach areas and the artificial rocks in order to independently reclaim maritime social-recreational uses (Figs. 11.9 and 11.10).

From Mergellina, it is possible to reach the Posillipo hill, characterised by a high quality of landscape. This is the most evocative stretch of coastline of the Naples urban amphibious due to the positive relationship established between urban naturalness, considering the vegetation on the side of the hill facing the sea, and the significant historical elements, from the Virgiliano Park to the Pausylipon Park, the Seiano Cave and the Gaiola underwater park; however, here the accessibility to the city-sea interface is extremely limited, due to both the complex orography and the coastal privatisation by private users who have incorporated accesses to the sea into their residential or tourist properties, not always in a legal way (Fig. 11.11).

Fig. 11.9 The Molosiglio gardens act as a green buffer zone between the port and the urban area of the coastal interface (picture by the author)

Fig. 11.10 The promenade of the Mergellina area is one of the main walkway of the Neapolitan city, with its recreational, cultural and ecological values (picture by the author)

Fig. 11.11 The Posillipo hill is an actual littoral green lung; however, accessibility to the coast is highly unequal (picture by the author)

This opens the discussion to the issue of maritime concessions and management of the coastal maritime domain. The shore is high and often rocky except for few small inlets that unfortunately only allow access to residents. Mobility is strongly compromised by an infrastructural system composed of narrow roads with no parking areas. Thus, although Posillipo constitutes the portion of shore with the greatest landscape

Fig. 11.12 Marechiaro area, in the Posillipo zone, hosts one of the few bathing areas freely usable by Neapolitan users, even if its accessibility is not equally easy for all of them due to the physical conformation of the site (picture by the author)

value and with the better water quality for bathing of the city, it can be said that it is not equally accessible for the littoral community. It can be considered a space with a marked environmental character that is mainly reserved for private owners, usually belonging to the wealthier classes of the population. Indeed, this stretch of coast is characterised by the presence of private residences, some of them possessing historical and exceptional landscape value, and restaurants or accommodation facilities that prevent easy access from both land and sea. In addition, many areas are poorly maintained and present high risks of landslides, combined with widespread illegal building (Fig. 11.12).

Finally, in continuity with Posillipo, there is the last stretch of the Neapolitan urban coastline, namely Bagnoli: as mentioned earlier, this is a former industrial area where social conflicts are linked to the desire for recreational and bathing uses of local sandy shores, but also for new public spaces and tertiary and sports poles in a peripheral part of the shore that has been often left at the mercy of itself by public institutions. Particularly important is the issue of recovery for land and sea areas, since this place is deeply affected by heavy metal pollution due to the former productive activities, as well as the question of public land ownership: this elements should be well considered by scholars and technicians for the purposes of careful and accurate planning process. In this regard, the last ideas competition for the Bagnoli coast, managed by the government agency Invitalia, was concluded in June 2021, proposing the realisation of recreational and social facilities, as well as the definition of beaches and bathing spots, next to a large natural park: however, there is still scepticism about the actual implementation of this proposal (Fig. 11.13).

Fig. 11.13 The large abandoned productive area of Bagnoli, with its famous panoramic pier. The potential of this place must also be read in a landscape key (picture by the author)

References

Bertoncello B, Dubois J (2010) Marseille Euroméditerranée, accélérateur de métropole. Parenthèses, Marseille

Bertoncello B, Hagel Z (2016) Marseille: une relecture de l'interface ville-port au prisme de l'habiter. Vertigo 16(3). OpenEdition Journals

Bertoncello B, Rodrigues-Malta R (2003) Marseille versus Euroméditerranée. Annales de géographie 632: 424–436

Bertrand R (2012) Histoire d'une ville: Marseille. Canopé – CRDP, Neuville-de-Poitou

Consalès JN, Goiffon M, Barthélémy C (2012) Entre aménagement du paysage et ménagement de la nature à Marseille, la trame verte à l'épreuve du local. Développement durable et territoires 3(2)

Deboudt P (ed) (2010) Inégalités écologiques, territoires littoraux & développement durable. Presses universitaires du Septentrion, Villeneuve-d'Ascq

Palmentieri S (2016) Innovazione e ridisegno degli spazi urbani: dai vuoti ai poli di sviluppo. L'Area Est di Napoli. In De Falco S (ed) Innovazione, competitività e sviluppo nei territori dell'Unione Europea. Edicampus edizioni, Rome, p. 109–122

Papa R (ed) (2010) Napoli 2011. Città in trasformazione. Electa, Naples

Zalio P (1996) Une histoire du milieu marin méditerranéen. Le cas du golfe de Marseille (XVIIIe - XIXe siècles). Enquête 4: 191–210

Chapter 12
Normative Analysis of Existing Urban Plans and Policies for the City-Sea Interfaces of Marseille and Naples

Abstract From a normative point of view, the management models of the urban coastal spaces of Marseille and Naples denote a conspicuous difference, also because of the different level of implementation of the governance dynamics in these two littoral interfaces. The main policies and plans related to the development of the urban shore, in a direct and indirect way, are therefore highlighted in order to understand the priority addresses and the real involvement of the institutions in the management of this complex spatial entity, starting from the beginning of the new millennium.

Keywords Coastal management · Normative analysis · Legal comparison · Marseille urban shore · Naples urban shore · Coastal development

12.1 Coastal Planning in Marseille as an Interpretation of Current Social Needs

Current legislation in Marseille takes great care of the socio-recreational value of the coastline. As stated above, the last thirty years of the twentieth century have seen an enhancement of the urban shore in favour of sociality, with the realisation of the Pointe Rouge tourist harbour, the city purification plant and the Parc Balnéaire du Prado, in the Rade Sud, and the Corbière beach, in the Rade Nord, alongside an overall improvement of bathing water quality. From 2000 to 2010, the municipality increasingly included users in its coastal policy, promoting the *Plan de Gestion de la Rade de Marseille* (2009). This policy instrument proposes approximately seventy actions in favour of the sea and the coast of Marseille, providing for the implementation of a better and more equal accessibility to the urban littoral, with a focus also on the removal of architectural barriers for users with reduced mobility. To this is added the enhancement of leisure docks, an in-depth analysis of the criticalities of the numerous urban beaches along the Marseille city-sea interface and a redevelopment of coastal pedestrian paths. Nevertheless, the growing weight of issues related to social uses, coastal pollution and environmental hazards led to further urban planning and normative elaboration in the 2010–2020 decade: five additional documents

I. Pistone, *Urban Coasts in Socio-ecological Transition*,
https://doi.org/10.1007/978-3-031-70783-4_12

Fig. 12.1 The city-sea interface of Marseille from Notre-Dame de la Garde Marseille (picture by the author)

regulating the use of the Marseille city-sea interface, from different points of view, were thus launched (Fig. 12.1).

The *Plan Plages et Littoral* (2010) has a strong socio-environmental character and is part of the municipal strategy to protect and develop Marseille coastal areas. First of all, this instrument aims to improve the quality of urban waters for recreational and bathing purposes; it also plans to protect areas available for social and bathing activities from coastal erosion, reinforcing the existing defence structures against risks from the sea and building new ones where necessary. The policy aims to improve the general conditions of urban beaches, from the point of view of cleanliness and hygiene, but also to increase the perception of safety of coastal areas through measures to safeguard bathers; it provides better services and diversifying the range of functions connected to urban beaches, also through the integration of appropriate and responsible commercial activities. Finally, special attention is paid to the use and accessibility of these public places, in relation to users with reduced mobility, and to the efficient connection between the city and its maritime-recreational areas, also through cycle paths and public transport services. The *Plan Plages et Littoral* shows interest in the social aspect of the city-sea interface, proposing the realisation of spaces intended for open-air sports and service facilities (such as bathing-huts and toilets); besides, it supports the creation of new beaches and redevelopment and enlargement of the existing ones, especially along the Rade Sud, in order to meet the social needs of citizens, who must become aware of the sustainable use, protection and risks of the coastal space.

The subsequent *Plan Nautisme et Plongée* (2011) implements the nautical character of the city through tourism initiatives, improving the quality of leisure docks, proposing a rational and sustainable use of water spaces, and involving the population in activities related to recreational boating and amateur diving. The Marseille government is showing its intention to avoid for sea planning the same mistakes made over the years on land, where the available public space has been eroded by massive road infrastructures and parking lots, frequently resulting in hardly reversible interventions. Similarly, excessively increasing the capacity of port facilities, as well as building new berths without an accurate assessment of the inevitable impacts, can be incredibly damaging to the landscape, in relation to the image that the city offers from the sea and the vulnerability of the surrounding natural coastal environments. The actions envisaged in the document therefore respond to several issues, aiming firstly at the sustainable development of leisure harbours, rationally increasing their capacity and developing related services, and adapting the coastal interface and maritime logistical nodes to underwater recreational and cultural activities. Efforts are also made to stimulate nautical and diving-related tourism by supporting major national and international sea-related events, but also by involving urban coastal society in these initiatives and promoting synergy with sports federations, associations, research centres and universities. For these reasons, the ports of Saumaty and L'Estaque, in the Rade Nord, and the port of Frioul have been developed; besides, it has been created a centre of excellence dedicated to boating and diving, also in the Rade Nord.

The drafting of the *Plan de valorisation du milieu marin et de ses ressources* (2011) is linked to a more environmental and ecological aspect, consolidating cooperation between public and research institutions with associations and ordinary users of urban coastal society to limit maritime degradation. Among its primary objectives, there is the identification of critical economic, social and ecological issues related to the city's marine resources. For this reason, the document supports scientific research in coastal areas, promoting projects for restoring degraded marine areas and the construction of artificial rocks for defence against environmental risks and coastal ecosystem development. It aims to develop social education and awareness-raising actions in order to operate an integrated, community-based and sustainable management of the natural system at the base of the coastal interface, involving expert and non-expert users in the ecological planning of the coastline.

Similarly, the *Plan de Gestion et de Valorisation des Espaces Naturels Littoraux et Insulaires* (2013) seeks to valorise the natural spaces of the Marseille coast as well as its islands, with a view to ecological transition, by stimulating awareness of maritime issues among users. This policy is articulated in a differentiated manner according to the site of interest along the Marseille shore, through specific management schemes, operational projects, but also the creation of protected natural areas where necessary. Generally speaking, however, it is possible to identify common objectives for each action in the document. First of all, it is stated that an in-depth knowledge of the coastline, islands and maritime resources of Marseille is essential for sustainable development and efficient management of the different uses of the city-sea interface: a special policy is therefore proposed in order to prevent damages that may alter the ecosystem status, in relation to urban waters but also to the urban greenery along the coastline (preventing fires, for example), taking into account the ecological transition of human activities at sea and on the land. Once again, emphasis is placed on adequate social education on the environmental aspects of the city-sea interface and on good practices to be developed in this sensitive area of the city.

The *Contrat de Baie*[1] is another document with great value for the Marseille shore development: it proposes a strategic vision for the coastal territory, both from an ecological perspective and with a view to the management of social uses. This document has been drafted through the joint efforts of the municipality, the province and research and urban planning bodies. It stems from the will of local actors to improve the quality of bathing water and the littoral *milieu*, proposing solutions mainly on a small scale. Among its objectives, there is the reduction of marine pollution levels also for recreational purposes, the protection of the ecological quality of the coast, as well as the organisation of a more efficient coastal governance in relation to local uses and stakeholders: the *Contrat de Baie* aims, in fact, to systematise all coastal plans, programmes and actions with the same purposes. Moreover, since the coastal area analysed by the document often presents conflicting economic, tourist, cultural and recreational stakes, community awareness and involvement are sought, with the establishment of partnerships with competent authorities regarding bathing activity and beaches and the creation of a permanent observatory with the purpose of providing information support and developing databases dedicated to the coastal interface. In addition, it is proposed to regulate bathing functions for sandy or rocky beaches serving the majority of the coastal society, as well as for small beaches used mainly by residents. It is noted that accessibility to these places is not always easy, especially by land: the risks that afflict these crucial points of the coast are therefore studied, with the possibility of creating new coves for access to the sea by boats.[2] From an ecological point of view, the *Contrat de Baie* identifies littoral areas subject to the greatest environmental criticalities, such as erosion and flooding, in relation to criteria such as the resident population, the surrounding habitat, the presence of buildings and working activities: these points along the shore are called *Territoire à Risques important d'Inundation* (TRI). From an operational point of view, the priority project actions are identified, in agreement with citizens and authorities in charge, assessing their feasibility and urgency, also on the basis of their location on the Marseilles territory; the plan then provides for the finalisation of the interventions through the elaboration of the specific objectives and the allocation of the necessary economic budget. In essence, the document has a strong socio-environmental value,

[1] The period of validity of the *Contrat de Baie* runs from 2015 to 2022. Subsequently, the *Contrat de Tranisition* replaces it for the period 2022–2024, assessing the impacts of its implemented actions yet maintaining the same strategic line.

[2] This reasoning about accessibility is very relevant for the management of the sea in Marseille, since there is currently only one cove for access to the sea for boats not affiliated to nautical clubs. The so-called *Cale de la Lave is* located in the Rade Nord and is a very congested gateway to the urban water.

seeking to restore the ecological quality of the interface, facing the issue of land consumption and managing ecosystem services, studying the use of public coastal areas and limiting the impact of port infrastructures on beaches and coastal spaces dedicated to users, through ecological planning techniques and an accurate management of the functions of the coastal strip that are most closely connected to the sea.

At the metropolitan scale, the *Plan Local d'Urbanisme Intercommunal* (PLUi), finally approved in 2021, focuses on the well-being of coastal areas and their recreational and environmental potential. Compatible with the *Loi Littoral*,[3] it highlights different types of coastal spaces to be organised and protected, enhancing the ecological and functional quality of green areas along the city's coastline through an ecosystem approach. In particular, *spaces close to the shore* are highlighted, imposing strong limitations to building: these areas include urban spaces between the city and the sea, where there are clearly conflicts linked to the contrast of the two systems; they are influenced by the surrounding urbanisation and their delimitation may also be indicated by the presence of road infrastructures (in accordance with the maritime city edge notion). Instead, *valuable natural spaces* are areas that require special care to preserve their landscape and environmental features, while *urbanisation fracture spaces* call for even more scrupulous protection of the ecological functions and ecosystem services within the littoral zone: a coastal buffer zone of one hundred metres from the shore is applied to these spaces. At the same time, the PLUi proposes a recomposition of the functional and spatial elements of the Rade Nord, reconfirming the social and bathing characteristics of the Rade Sud, through interventions aimed at reconnecting the city, port and sea systems, but also at dedicating urban spaces to satisfy the current social demand. The objective is the enhancement of the city-sea interface by favouring soft mobility and intervening on the transition areas between the urban and maritime systems with the integration of urban parks and green–blue public areas: public space is considered as an extension of the urban beaches, ensuring greater visual permeability towards the sea. The Marseille Rade is thus planned to host new socio-recreational spaces with high ecological quality in the northern part, functions linked to the productive and logistical context in the middle area, and actions for spatial and functional reconnection of coastal fragments in the southern zone, which is already devoted to community uses of the sea.

These indications recall the addresses defined within Marseille municipal urban planning by the *Plan Local d'Urbanisme* (PLU), approved in 2013 and subsequently integrated in 2015. According to this plan, the valorisation of coastal areas can be achieved through the integrated management of public places, ensuring a better spatial, social and environmental quality and enhancing both direct and indirect accessibility. It proposes the realisation of recreational and cultural poles in relation to the sea, especially in the northern part of the city, through the reconversion of former productive sites and interventions on road connections and soft mobility,

[3] The *Loi relative à l'aménagement, la protection et la mise en valeur du littoral,* simply known as the *Loi Littoral,* was approved on January 3rd, 1986: it has the task of regulating and protecting French coastal development, also managing free and equal accessibility to the sea resource.

in a logic of preservation of environmental values: the Rade Nord has deep-rooted economic stakes, due to the presence of the port infrastructure and several large productive-commercial activities, but it still offers diversified spaces that can be developed towards a recreational value, respecting the social demands of local users. At the same time, the plan is directed towards a better management of the public coastal spaces and urban beaches along the entire city-sea interface, indicating actions aimed at improving social fruition and the existing facilities, expanding the functional offer for community and recreational uses and thus favouring the opening of the city towards the sea. Reference is also made to the improvement of the network of coastal green–blue spaces, in order to maintain a high level of protection for natural spaces while adapting to environmental challenges and the social uses they can host, also in relation to water quality.

With regard to the planning of the city-sea interface of Marseille, reference must finally be made to the *Charte Ville-Port* (2013), drawn up by the Port Authority of the *Grand Port Maritime de Marseille* (GPMM): in particular, the port characterises the entire northern part of the city, even though it influences prominently and not always positively the relations with the surrounding spaces; for this reason, this strategic document has been realised, with a time horizon of fifteen years, trying to revisit the city-port interactions in relation to the dynamism of the harbour spaces and the development of the urban territory, integrating economic-financial but also social balances. The main intentions of the *Charte Ville-Port* are certainly related to the development of port activities that should be included in urban projects,[4] reaffirming the productive and logistical vocation of coastal water basins: however, it recognises that the emergence of a strong city-sea relationship, in which the port infrastructure is a major element, requires to have a mixed functional planning between the harbour function and social uses. For this reason, it declares the intention to partly redirect the ship traffic in order to favour civil coastal use. From a spatial point of view, the strategic document divides the coastal interface into three macro-environments, in relation to the presence of the port. The southern zone hosts the passenger port and is in close contact with the urban centre, becoming a potential urban hinge and a public space of great social quality: in fact, buildings for public and socio-cultural have been realised there, such as the MuCEM, the CEREM and the multifunctional centre *Les Terrasses du Port*; around these elements, maritime squares and *boulevards* have been developed. The middle zone stretches from the Arenc area to the border with the *16th arrondissement*: here the port function is confirmed, while referring to the need to diversify the functional offer, cautiously considering recreational uses of the coast inside the harbour perimeter, such as leisure boating, and even the possibility to partially allow free passage of pedestrians. Finally, the northern zone is indicated as suitable for the creation of recreational areas and public proximity spaces, reaffirming the social-tourist qualities of the area: new economic activities are proposed for the Saumaty area (the northern portion of the Marseille port), like the redevelopment of existing harbour structures to reduce the visual and environmental impact

[4] According to Law 2008–660, coastal basins, their land areas and docks are the property of the port, which takes care of their development and is in charge of enhancing them.

and the functional reconversion of some areas, improving the offer of services to the community; the need to extend the surface dedicated to urban beaches is also identified to favour equal accessibility to the sea. The *Charte Ville-Port* mentions actors mainly from the logistics, commercial and financial sectors, but also identifies ways to include community participation in coastal planning. However, while promoting a planning synergy with society, the interventions proposed and implemented have been predominantly managed by institutional consultation, dedicating the central area of the coast, several kilometres long, to logistics and port activity.

For this reason, the *Dialogue Ville-Port*[5] have been set up: they are cycles of meetings and participatory activities with public bodies, technicians and representatives of the population that, starting in 2019, channelled the request of stakeholders to be informed and involved in the planning of the coast and port areas, which have such an impact on the lives of citizens: through thematic workshops, the participants produce transformation hypotheses for the city. Interesting, for example, is the work carried out for opening to the public the *digue du large*, a walkway approximately 7 km long facing the GPMM that protects the urban coastline from environmental risks, even if it is proposed as a maritime promenade that would strongly define the social and recreational relationship between the urban coastal society and the sea of their city. The propensity of the Marseille case study to the socio-environmental transformation of the urban amphibious is thus clear.

12.2 Governance Conflicts and Weak Social Planning Along the Coast of Naples

The coast of Naples is an interesting case, as the quality of the water, the strong logistic-port activity and the privatisation of littoral accesses preclude adequate social uses, also due to the lack of virtuous effects of the industrial decommissioning, exacerbating the social tension regarding the use of the maritime landscape on a neighbourhood scale. This is witnessed by the inhabitants of the eastern area of San Giovanni a Teduccio, where there is a strong desire to enjoy the local shore, despite the poor ecological condition of the water and the urban beach, and the community of Bagnoli, at the western end of the city, which is divided between the desire to establish social and recreational uses along the shoreline and the difficulties of a peripheral area affected by the criticality of coastal recovery and land ownership, often forgotten by institutions. Both areas have been classified as *Siti di Interesse Nazionale* (SIN) due to their complex environmental condition: despite this, they are still unresolved issues within the Neapolitan coastal landscape (Fig. 12.2).

---

[5] The *Dialogue Ville-Port* initiative, established in 2019 and renewed in 2021, seeks for the integration of residents' expectations within urban governance processes, the promotion of sustainable relations between port activity, related coastal urban areas and social uses of the shore, and finally the design and implementation of integrated urban maritime projects.

Fig. 12.2 The city-sea interface of Naples from Sant'Elmo Castle (picture by the author)

Neapolitan planning does not offer many particular solutions to increase the landscape and social value of the urban coastline. At the municipal level, the *Preliminare di Piano Urbanistico Comunale* (PUC), approved in 2020, proposes limited interventions for non-port areas along the shore. It reflects on the issue of accessibility, emphasising the relationship between port and city, especially in the eastern part of the littoral zone. It is proposed the implementation of a specific regulation for the coastline, in order to guarantee community accessibility to the sea and protect the coastal landscape, redefining the relationship between the waterfront and the historic city in a qualitative manner: the objective is the functional and environmental improvement of the city-port transition zones, in competition with the Port Authority; however, there is a lack of in-depth consideration of the theme of coastal recreation and sociality, as well as the localised planning of proximity public spaces that can constitute a coastal green–blue network. Reference is also made to the importance of social participation in the design of public space and socio-environmental regeneration, although no particular strategic guidelines are identified. From the environmental point of view, the document identifies some areas of the Neapolitan city-sea interface subject to the risk of coastal erosion, in particular the Posillipo stretch of coastline: it indicates the need to propose interventions, even localised ones, to defend the shore from the dangers of the sea, defining implementation methods and legal limits. This critical issue is in fact linked to the community's enjoyment of the water resource. The *Preliminare di PUC* therefore identifies the need to regulate the access points to the sea, providing the necessary indications for bathing establishments along the coast: for this reason, it proposes to identify the territorial areas whose transformation is subordinate to the drafting of specific implementation plans, with the indication of the appropriate urban planning parameters, as well as the types of light and non-impacting actions that may be allowed within the process of socio-functional reconnection between the city and the sea. The intention is to give the possibility of setting up temporary facilities for bathing or other recreational uses, even outside the summer season, postponing to a public selection process the choice of the related managers who will have to sign a special agreement, although the specific procedure is still undefined: among these hypothetical installations, light boardwalks and mobile jetties for the temporary mooring of boats are included. Despite some positive intentions, there is still a lack of clear definition of the implementation modalities for these uses of the coast, albeit the *Preliminare di PUC* is merely a strategic document. In this sense, it is worth mentioning also the recent document *Per una città giusta, sostenibile e attrattiva*, approved in March 2024, which broadens the strategic horizons of the above-mentioned *Preliminare*. In particular, it defines the urban sea as a public space to be valorised through actions for improving accessibility and promoting cultural initiatives; at the same time, it recognises the role of the green-blue infrastructure in reducing environmental risks linked to climate change, hypothesising measures for adapting to the phenomena of coastal erosion and sea-level rise that afflict the littoral interface. In addition to this, reference is made to the dynamic spatial dimension of the coastal-maritime landscape, emphasising the need for some priority interventions, related to the widening of public beaches through nourishment (especially in Bagnoli and San Giovanni a Teduccio, in relation to the regeneration of local former industrial sites and degraded open green areas), the enhancement of accesses to the sea in the Posillipo area and the redesigning of the Neapolitan littoral promenade in function of a better use of the city-sea interface; the document also declares the intention to mediate the complex relationship between public urban areas and port spaces. Even in this case, however, the document can be placed only within the municipal strategic framework, even though it highlights issues of great interest for the socio-environmental development of the Neapolitan urban amphibious.

On the other hand, the current *Variante del Piano Regolatore Generale* (PRG) (2004) has already shown a discrete operational attention to social aspects, environmental sustainability and landscape protection. The *Variante* emphasises that the urban coastline of Naples is characterised by a vast extension, but less than one third of its length has a real direct relationship between the city and the sea: this is due to the fact that almost five kilometres of the coastal interface are in fact occupied by the massive port infrastructure; another critical element, in the eastern area of the urban settlement, is the presence of the railway, which acts as a fracture between the urban system and the maritime system. Although the harbour historically constitutes a cornerstone of the economic structure of the city, it must nevertheless be taken into consideration that it currently represents an obstacle rather than a point of access to the sea, operating a socio-cultural rift in the urban amphibious: the document defines the port system as a territorial sphere that is independent from the rest of the city, with a varied range of productive, logistical, commercial and tourist functions, but little social permeability. In this sense, the issue of existing docks and berths for leisure boating is also discussed. About twenty years ago, the demand for boat moorings was already strongly unsatisfied and steadily increasing: the plan therefore seeks to increase the infrastructure for pleasure sailing, enhancing the sector's potential, even though the strong constraint of lacking space along the city-sea interface still persists. The relevance of this issue is reflected in the need to limit the environmental degradation generated by the irrational and irresponsible location of berthing places, which are often illegal. The intention expressed by the *Variante* is therefore to regulate the shore in order to favour an efficient integration between the city and the sea, reducing socio-ecological impacts to a minimum. Strategically, reference is made to the easternmost area of the urban coastal strip, bordering the Neapolitan port, namely the area between Vigliena and Pietrarsa, where it would be possible to enhance the natural aspect of the city-sea interface, improving littoral usability and accessibility for local users, by locating high-quality social functions: hence, it is proposed to create a large linear green–blue park following the coastline starting from San Giovanni a Teduccio, overcoming municipal boundaries and connecting together the various coastal public spaces of the Neapolitan metropolitan area, favouring pedestrian and bicycle mobility and the strong landscape values of the territorial littoral at a larger scale; however, it remains an unrealised project until today. From the environmental and ecological point of view, the plan seeks to actively protect the Neapolitan environment, fostering its valorisation in a dynamic and non-binding manner; with regard to coastal risks, it identifies the criticality of landslides affecting the parts of the city-sea interface characterised by high vertical fronts, which are now fragile due to past mining activity.

Currently, the municipality identifies twenty-four urban beaches[6] (according to the *Mayoral Ordinance no. 248 of April 22nd, 2021*) located in three areas of the Neapolitan interface: the bathing area of the City Centre extends from the Molosiglio cliff to the foot of the Posillipo hill, crossing the promenade of Via Caracciolo and Mergellina; the bathing area of Posillipo-Marechiaro stretches from Mergellina to the Trentaremi Bay, including important elements of the city's architectural, archaeological and natural heritage; finally, the bathing area of Bagnoli-Coroglio includes the western part of the urban coast of Naples as far as the border with the municipality of Pozzuoli. In this classification, it is not mentioned the San Giovanni a Teduccio area, where the ecological quality of the water is extremely critical, even if social demand for direct contact with the sea resource still persists. Moreover, fifteen of the twenty-four urban beaches identified are private: only three of the remaining nine free beaches, are identified as equipped with bathing services, but they are not always easily accessible. One example is the beach of the Underwater Park of Gaiola. This area is regulated by the *Interministerial Decree of August 7th, 2002*: according to the legal framework, the site is divided into a zone A of integral protection, in which

[6] The municipality of Naples has mapped the urban bathing system in 2021. For more information, please visit the following website: https://www.comune.napoli.it/flex/cm/pages/ServeBLOB.php/L/IT/IDPagina/12668.

it is not possible to adopt or renew any measure for the use of the maritime and coastal domain unless for safety, service or research reasons, and a zone B, in which it is possible to implement indications for social and cultural activities, in agreement with the competent authorities, but always with respect to the protection of the Park. The area may be subject to restrictions for the access to bathing areas, in relation to the naturalness of some zones, but also for other reasons: for example, during the past pandemic period, the fruition of the beach zone was severely restricted due to the health emergency. Despite the fact that it is a free beach, bathing is forbidden in zone A, except for special reasons of study or research, while in zone B it may be permitted, albeit with limits relating to the prescriptions of the managing authority and the rules of the port captaincy: the same limitations apply to nautical activities in the water of the protected area.

It is thus evident from the municipal mapping that swimming is in various cases not allowed, while most urban beaches are privatised, affected by polluted waters or poorly equipped for bathing activities, intensifying the problem of unequal accessibility to the urban shoreline: for example, in the area comprehending the historic urban fabric of Santa Lucia, the Riviera di Chiaia and the Mergellina promenade, despite the tourist-recreational matrix, the coast lacks adequate bathing facilities; in the Posillipo area, the complex orography reduces the possibility to have an easy fruition of the sea, while local inhabitants and economic activities monopolise most of the coastal accesses, as mentioned above. Besides, Posillipo is regulated by the *Piano Territoriale Paesistico dell'area di Posillipo.* This plan states that, within a buffer zone of three hundred metres from the coastline, interventions for the conservation and redevelopment of the existing natural and vegetal system are permitted, while the compatibility of the facilities related to beaches and bathing establishments with the local landscape, environmental and archaeological values must be verified before any project authorisation: only hygienic-technological adaptation is allowed, without further increases in volume. The same indications apply to commercial and restaurant buildings and existing nautical clubs. It is also emphasised that the landscape-environmental redevelopment of the area is a municipal responsibility, as well as the identification and removal of unauthorised built elements that disturb the enjoyment of the coastal area. However, still remains a general difficulty in accessing the coastal recreational areas of this relevant portion of the city-sea interface.

Moreover, the Posillipo cliff is the area of the urban shore that is most affected by the environmental hazard. According to the analysis of the *Piano per la Difesa delle Coste* of the Basin Authority, the area appears to be subject to a very high risk in terms of coastal erosion and flooding, as well as to the risk of landslides: this clearly has a very strong impact on the usability of the coast and the sea, further limiting the possibilities of access to the city's water resources and placing precise safety constraints on any future planning and design intentions for this littoral space. Other areas of the urban amphibious where the document highlights high risk of flooding and erosion are the urban beaches of the Rotonda Diaz, in the heart of the urban shore in the Riviera di Chiaia, the areas related to private bathing establishments at the base of the Posillipo hill and the coast of San Giovanni a Teduccio, close to the port infrastructure. According to the aforementioned cartographic sources, the sea defence works, essentially limited to traditional artificial rocks parallel to the coastline, are subject to the risk of overflowing, especially in the area of the Riviera di Chiaia: this indicates an essential lack of efficient defence works against maritime dangers.

It should also be emphasised that coastal management is affected by an unclear demarcation of coastal state property between the municipal administration and the Port Authority: in fact, not only areas related to the logistics function fall under port management, but also a substantial part of the entire Neapolitan coastline. This governance anomaly weakens the municipality's interest in enhancing coastal public spaces. From a planning point of view, not only the aforementioned *Variante del Piano Regolatore Generale* insists on the port of Naples, but above all the *Piano Regolatore di Sistema Portuale* (PRSP), now in phase of drafting in according with the new sectoral legislation (Decree Law 169/2016). With the introduction of the new legislative decree, the old *Piano Regolatore Portuale* (PRP) of the Naples harbour, drafted in accordance with the Law No. 84/94, adopted by the Port Committee in 2012 but never officially approved, no longer serves any purpose. Port planning therefore follows the indications of a specific strategic document, namely the *Masterplan del Porto di Napoli* (2017). This document analyses the criticalities of the management model for the harbour and the coast of Naples: for this reason, it is considered appropriate to reorganise the system of concessions in the port area, in relation to the different functions that may be present inside the port or in the bordering spaces, also in the perspective of future changes and evolutions. The intention to intervene on the western area of the port interface (San Vincenzo Pier) from a tourist-social point of view is also highlighted, in order to balance the relationship between the harbour and the city: in fact, the historic port should be a social hinge, not only because of its economic value, but also because of the presence of public and cultural green spaces. In any case, it should be noted that the Port Authority and the Municipality of Naples have recently identified the conditions for a clearer definition of the planning competences of each of the two institutions, although this agreement has not yet had an evident effect on the socio-environmental transformation of the urban amphibious.

On a broader scale, the *Piano Territoriale di Coordinamento* identifies the ports of the province main cities as the major nodal elements of the coastal territory, albeit without proposing a strategic cohesion of harbour areas between economic-logistic functions and social potential. Another relevant element of the legal framework in Naples is the *Piano di Utilizzazione delle Aree Demaniali* (PUAD), a regional planning instrument approved on December 20th, 2022, that, despite its broad range of action, represents a directive specifically designed for recreational-tourist purposes of the coast. Even if it is still in a drafting phase before its final approval, the PUAD identifies three coastal settlement categories (*A. High tourist value, B1. Ordinary tourist value, B2. Low tourist value*) in relation to the endowment of recreational-recreational services, water quality and cultural-landscape value. In particular, coefficients are applied to assess the recreational quality of the shore, in relation to the ratio of tourists to residents, the available concession areas, the presence of parks and accommodation facilities and bathing functions: the latter parameter is calculated on the basis of environmental quality criteria and water monitoring by ARPAC (*Agenzia*

Regionale per la Protezione Ambientale in Campania). However, it is surprising that the Naples coast is placed in the middle category B1, as it does not fully meet the abovementioned parameters. From a management point of view, the PUAD indicates that state-owned areas for tourism and recreational purposes have to be managed by the municipalities in which they are located, with the exclusion of those falling within ports of regional or inter-regional importance, managed by the Region itself, or areas under the control of the Port Authority. In order to plan the bathing functions along the coastal space, explicit reference is made to the concepts of free sandy shore, namely a strip at least 5 m deep, calculated from the foreshore, of shaded resting areas and of beach services that bathing-related spaces must be equipped with. Moreover, it is stated that at least 30% of the total length of sandy shores, as well as 30% of the total surface of other kinds of spaces within the maritime public domain, has to be granted to the coastal community for socio-recreational activities. It is also established that urban centres along the coast should locate at least one area within their municipal perimeter with peculiar landscape, environmental and bathing features to be destined as a free beach equipped with facilities and public services for the recreational enjoyment of the sea.

On a local scale, the PUAD is implemented by means of the *Piano Attuativo di Utilizzazione delle Aree del Demanio marittimo* (PAD), identifying intervention areas with design, environmental, ecological and landscape indications, pointing out the soft mobility system and the accesses to the sea, but also eventual coastal roads, the presence of equipment or specific entrances for users with reduced mobility, as well as technological networks; furthermore, the PAD describes the current use of maritime domain areas, distinguishing free beaches, beach services and bathing establishments (also with other social functions, outside the summer season). The PAD must also guarantee that services connected to bathing and commercial or catering activities must not in any way affect the landscape value of the coastal interface, avoiding barriers or obstacles to views towards the sea: it is in fact established that boundary fences must not exceed a height of 1.5 m and must be built with materials compatible with the surrounding environment and necessarily interrupted at 5 m from the shoreline. In order to meet the planning indications from the PUAD about the minimum surface of free fruition of sandy shores and other coastal state-owned areas for social, recreational and bathing purposes, public accesses to the sea should be built at least every 200 m, with a view to equal use of the urban coastline. The city of Naples has not drafted this operational plan yet, since the elaboration of the PUAD is still ongoing, but it is interesting in relation to the criticality of local governance: the PAD in fact describes the criteria for the allocation of maritime public concessions that should convey a qualitative improvement of the coastal areas for which they are granted, from environmental, landscape, functional and accessibility points of view, focusing on possible recreational uses. The assignment of licences must also be in line with the reference PAD: this factor justifies even more the need to promote this very important urban planning tool for the management of the city-sea interface. However, both the PUAD and the PAD show evident lacks in the definition of the social uses of the coastal resource and in the protection of its environmental aspects: the question of defence against climate risks is not dealt with in depth and no indicators are expressed to evaluate the efficiency of public areas; on the contrary, the emphasis is placed on issues related to the economic profit of bathing establishments. Although the economic sphere is relevant in the management of the city-sea interface, a more socio-environmentally oriented approach would be necessary for the valorisation of the urban littoral spaces.

12.3 Complexities and Differences in the Planning Frameworks of Marseille and Naples

In line with the objective of interpreting the needs and aspirations of coastal communities regarding coastal leisure and recreational spaces, the first phase of the comparison focused on planning and regulatory analysis, highlighting a clear difference between Marseille and Naples.

The French case study represents the model of European city that has been able to renew itself in the last decades between the old and the new millennium and to significantly change its image through phases of regeneration of disused coastal areas and urban waterfront, developing a specific knowledge about social activation for planning and co-design activities intended to read social needs and translate them into active projects. Starting from the first decade of the new millennium, however, a progressive attention to the theme of coastal planning emerges on the part of the local community and the competent institutions, perhaps in relation to the growing awareness of the importance of littoral interface management for socio-environmental and economic development. In this sense, Integrated Coastal Zone Management and Maritime Spatial Planning directives can be rightfully considered main influences among the European regulations, but also the presence of the *Politique Mer et Littoral* has been a source of planning inspiration. It can be noted a strong concern for the environmental quality of coastal areas dedicated to recreation, such as urban beaches, although frequency of visits is still not included in the criteria for drafting legal instruments. There is still a breach of duty towards the social demand for recreational fruition of the public coastal space: this is linked to a municipal poor attitude in listening to the population and, specifically, in public participation in coastal transformation projects: for instance, in the *Charte Ville-Port* the real social involvement was essentially bypassed by public institutions. Similarly, the study of the transition areas between the coastal-maritime system and the urban system still appears to be poorly investigated.

On the other hand, Naples pays the price of a delay in the implementation of the various plans and projects drawn up for the transformation of coastal areas. Despite the fact that social and co-designing workshops have been activated, in San Giovanni a Teduccio as in Bagnoli, going as far as feasibility project phase, none of these has been actually implemented, apart from small and limited interventions that are generally still in progress. From the planning point of view, there are considerably fewer documents than in the analysis of the French case. There may be several reasons for this. The apparent low interest on the part of the public institutions towards socio-environmental littoral issue may depend on a possible lack of planning competences at the local scale: this inevitably has negative effects on the production of adequate policies for the management and planning of the public coastal space. Moreover, Naples denotes a particularly fragmented legal framework: land uses tend to frequently overlap and governance boundaries are not always well defined. An example of this is the massive presence of the Port Authority in the management of the maritime domain, even outside the physical and functional harbour perimeter. The result is a gradual fading of public interests in the quality enhancement and design of urban beaches, small community spaces and parks along the shore. The interesting implication is that this management competition has led city planning to delve more deeply into the interactions between the city (especially in the port area) and its coastline. However, there is a general lack of attention to the social uses of the urban amphibious: there is a lack of documents specifically dedicated to the recreational dimension of urban coasts (differently from what emerged analysing the Marseille shore), while the town plan, namely *Piano Urbanistico Comunale*, that is still a strategic document, seems generally detached from coastal issues, especially in operational terms, even if it has been recently updated in its main objectives.

Moreover, the dynamics of concession-based relationships is still vague: indeed, it has to be noted that the planning tools intended to coordinate and regulate the features and quality of the interventions on the maritime domain are not yet operational, even if they are necessary for the equal social enjoyment of the sea resource. From the point of view of environmental risk, coastal erosion is one of the main criticalities: the *Piano Stralcio per l'Erosione Costiera*, drawn up in 2012 by the competent Basin Authority, locates the areas that are most exposed to the dangers of erosion and proposes possible risk mitigation solutions, even if its scale is wide (Fig. 12.3). Anyway, within ordinary planning, a real awareness of these negative factors, which can affect the life of urban coastal society, has not been developed yet: there are few references to operational actions for reducing the existing risks at the urban and local scale. It should be noted, in this regard, that both Marseille and Naples have been often burdened by flash floods or coastal storms in recent years: this issue should therefore play a prominent role within the planning and regulatory frameworks of the two cities.

In conclusion, it is possible to state that there is an evident disparity between the two analysed normative frameworks. Naples seems poorly involved in the social and recreational theme of its urban amphibious, while Marseilles denotes a greater propensity towards planning actions dedicated to bathing and leisure along the coast. Anyway, both cities are similarly lacking in implementing the principles of integrated coastal planning at the local level, although in both Italy and France there have been efforts to transpose the European directives, albeit in different ways and with different results: so, it is still missing a general framework that can effectively regulate the design of small spaces dedicated to urban coastal society, according to the principles of ecological planning. However, it must be emphasised that even from this point of view the Marseille system of coastal government is a step forward compared to the Neapolitan one, not only regarding the implementation of green–blue infrastructure:

TERRITORIAL | LOCAL				NATIONAL		EUROPE	
NAPLES		MARSEILLE		ITALY	FRANCE		
Projects	*Policies*	*Projects*	*Policies*				
		Corbière beach 1986 1988 Prado II 1995 Euromed		Legge Galasso 1985	1986 Loi Littoral	Rio Agreement (1992)	*1985-1995*
	1996 PTP Posillipo			L. Bassanini 1997	Code environnement (2000)	2000 European Landscape Convention	*1995-2000*
				L. 88/2001			
2002 PUE Bagnoli-Coroglio	Gaiola Park 2002					ICZM (2002)	*2000-2005*
2003 Città Scienza		2003 Plage Catalans = public		2004 Landscape Heritage Code			
Waterfront project 2004	PTC 2004 Variante PRG 2004						
Bagnoli jetty = pedestrian 2005							
Bagnoli ideas context 2006					CGPPP 2006	Bolkenstein dir. 2006	
		Euromed II 2007		2007 - Legge Finanziaria		2007 Flood Directive 2008 MSFD	*2005-2010*
2009 - PUA S. Giovanni			Plan Gestion Rade 2009				
	Piano Difesa Coste 2010		Plan Plages et Littoral 2010				
			2011 Plan milieu marin Plan Nautisme et Plongée 2011				
	PRP 2012		2012 Parc Calanques	D.L. 194/12			*2010-2015*
Re-opening Città Scienza 2013		2013 MUCEM (Capitale culturelle)	Charte Ville-Port 2013 PLU 2013 P. Espaces Littoraux et Insulaires 2013			EU Adaptation Strategy 2013 MSP 2014	
			Contrat de Baie 2015			Agenda 2030 (2015)	
	PRSP 2016	Pointe-Rouge project 2016			L. 2016-1087 (MSP - 2016)		
	2017 Port Masterplan			DPCM 01/12/17 (MSP) L. Bilancio 2019	Stratégie Mer et Littoral 2017		*2015-2020*
Bagnoli ideas context 2020	Preliminare PUC 2020					2020 EU Biodiversity Strategy	
			PLUi 2021				
	Mayor ordinance 248/2021 (urban beaches)						
2022 Piazza Municipio waterfont	PUAD 2022						*2020-Today*
	Mayor ordinance 232/2024 (urban beaches)		Contrat de transition 2023				
	Per una città giusta, sostenibile e attrattiva 2023						

Fig. 12.3 Multiscalar normative and planning benchmarking between Mareseille and Naples: partially or not all implemented actions are highlighted in red (elaboration of the author)

greater attention is also dedicated to environmental risks and to the defence of the city-sea interface from the wide range of dangers coming from the sea. Moreover, in both cities, the main planning operations have been started mainly with top-down initiative, even if social listening procedures have been activated. As a result, various interventions have mostly sought the regeneration of urban degraded areas and the redevelopment of public spaces for the purpose of commercial and tertiary real estate enhancement, even if they have not always realised the community expectations for the recreational fruition of beaches and urban parks along the coast. On the basis of the critical analysis of plans and policies drawn up for the littoral interfaces of Marseille and Naples with reference to the implemented transformations, it is stressed the weakness of co-designing activities: even if this kind of approach may give useful planning insights, in the context of the two analysed Euro-Mediterranean cities, the needs and desires of the coastal communities have been often limited within reduced areas by the political-administrative institutions, even though in some cases the latter have started virtuous processes, developing inclusive planning know-how.

Chapter 13
Geo-spatial Approach for the City-Sea Interfaces of Marseille and Naples

Abstract In this chapter, the geo-spatial approach developed for the city-sea interface is applied. In particular, the analysis of the main features of the urban coasts of Naples and Marseille refers to the maritime city edge portion of the two cities, that is, the area between the coastline and the first transit road parallel to the coast. This choice is supported by the presence of intense pressures in this area, from the point of view of the terrestrial sphere, directly influencing the water mirror and the functions it could take on, but also the innermost coastal areas of the city-sea interface.

Keywords Geo-spatial approach · Maritime city edge · Coastal public spaces · Coastal accessibility · Spatial planning · Maritime uses · Recreational uses

13.1 The Maritime City Edge as a Relevant Socio-Environmental Element for Coastal Planning

In the context of the spatial research of the main characters of the city-sea interfaces of Marseille and Naples, the previously described geo-spatial approach is used for the elaboration of a spatial data model that provides a functional description of the two coastal urban entities in comparison. According to the city-sea interface theory, this peculiar space hosts the most intense relationships between the maritime sphere and the city sphere: these coastal ecosystems include human and biophysical components and provide numerous benefits to coastal society, in the fields of tourism, recreation, fishing, commerce and aesthetic-cultural values; moreover, they are connected to numerous essential services such as energy production, environmental conservation and protection from natural hazards (Ranganathan 2007). It is therefore assumed that the maritime city edge, that is the coastal strip comprehended in between the shoreline and the transit road parallel to the coast, can therefore be the neuralgic core of these relations and impacts, which have both an environmental and social matrix: this part of the city-sea interface becomes a physical and functional hinge between the sea system and the city system, a space in which the ecological continuity is not broken by the main road infrastructure and in which the community comes into

I. Pistone, *Urban Coasts in Socio-ecological Transition*,
https://doi.org/10.1007/978-3-031-70783-4_13

contact with the city's water and at the same time enjoys the public coastal space. From a geographic and functional point of view, the maritime city edge thus assumes great value in understanding the entire city-sea interface; from a planning perspective, it becomes a key element in directing analytical studies towards a design resolution of littoral issues.

Indeed, the management of this coastal area must consider a wide range of complex, overlapping and often contradictory stakes. Decisions on regulatory priorities, such as design development initiatives, long-term plans or allocation of uses of coastal spaces, usually intertwine with changing social values and constantly evolving biophysical and functional contexts: balancing the needs of the different users of urban coastal society in relation to public spaces, private interests, ecological protection and development demands is a contemporary challenge that requires the combination of increasingly careful management plans and design actions (Loomis and Paterson 2014). For this reason, the geo-spatial approach here presented seeks to map the crucial elements of the maritime city edge, in order to understand which ones can bring enrichment to the urban shore and the littoral community, primarily from a socio-environmental and recreational point of view, in line with the objectives of this book, but also the eventual constraints to this kind of development.[1] This approach clearly recalls the identification of the benefits provided by ecosystem services, which are fundamental within decision-making processes oriented towards ecological planning: their value is linked to the well-being of society and the safeguard of the environmental status of places, motivated by the participation of the users themselves in the protection of ecosystems (Klain and Chan 2012).

With regard to the benchmarking context between the two Euro-Mediterranean case studies, the definition of the spatial data model involved the identification of five main layer categories: the *geo-physical structure* of the two cities, the *natural* and *built environments*, the *recreational uses* and the *soft mobility and accessibility* system.

The objective is to understand the composition of the maritime city edges of Marseille and Naples and to measure their different characteristics, in order to collect data necessary for careful planning of socio-recreational values and fair accessibility to the coast and the sea. For the study of the Marseille coastline, the base reference is the *BD Ortho 2020* image, while *Pléiades 2020* satellite images have been selected for Naples: the use of high-resolution and recent base maps has allowed to obtain the highest possible degree of precision in the digitisation phase of the various analysed zones.

[1] The elaboration of the following thematic maps, derived from the geo-spatial approach, has been carried out with the support of the CNRS ESPACE—UMR 7300.

13.2 Extent and Geo-Physical Structure of the Maritime City Edge

Within the geo-spatial analysis, the first difference to be found between the two case studies concerns the overall extent of the maritime urban edge holds from a dimensional and spatial point of view, despite the fact that the linear length of this element within the two cities is roughly similar (Fig. 13.1).

The municipality of Marseilles covers a total surface area of 238.4 km^2; its urban coastline extends from the municipal end of the 16th Arrondissement in the north to the Calanques national park in the south: functionally, it is possible to divide it into a northern portion with a mainly port function, a central portion with an urban and bathing character and a southern portion with an essentially naturalistic value, coinciding with the aforementioned protected park. The latter portion, however, is not included in the perimeter of the analysed maritime city edge because the Calanques area is essentially a space with a very low level of urbanisation in relationship to the other parts of the coastal settlement; besides, it is governed by peculiar regulations that limit design and socio-recreational flexibility for the purpose of protecting the surrounding natural environment. Consequently, the area studied in the geo-spatial approach corresponds to the first two stretches of the city's coastline, comprehending the Rade Nord and the Rade Sud, with a length of approximately 22 linear kilometres. In total, the maritime city edge of Marseille measures 5.20 km^2, corresponding to 2.18% of the total urban area. In the case of Naples, the municipal area has a smaller surface, with only 118.9 km^2. The studied coastline covers the entire municipal shore, from the western end of Bagnoli, on the border with Pozzuoli, to the littoral of San Giovanni a Teduccio, on the eastern edge of Naples. On the basis of these premises, the perimeter of the Neapolitan maritime city edge has been defined and its total size measures 7.5 km^2: from a functional point of view, it can be schematised into a western portion with a more industrial character (in relation with the former industrial area in Bagnoli) and naturalistic-residential character, a central area for mainly residential, tourist and recreational use and finally an eastern port and productive area. In percentage terms, the Neapolitan maritime city edge corresponds to 6.3% of the total urban area.

It is evident that although Marseille has a larger surface area than Naples, the French city devotes less space to the city-sea interface, at least in relation to its maritime city edge component. This is interesting because it strongly conditions the way in which spaces in direct contact with the urban sea are designed and indicates that the transit road, previously defined as the first strong interruption between the coastal and the urban spheres within the city, is in fact very close to the maritime ecosystem in the Marseilles case. On the contrary, Naples appears more characterised by its maritime city edge, at least from a dimensional point of view. The shape of this portion of city along the coast is certainly conditioned by the land reclamation interventions that municipalities may implement, but also by the presence of elements that spatially modify the urban shore. The database therefore investigates the features underlying the geo-physical structure that define the shape of the maritime city edge,

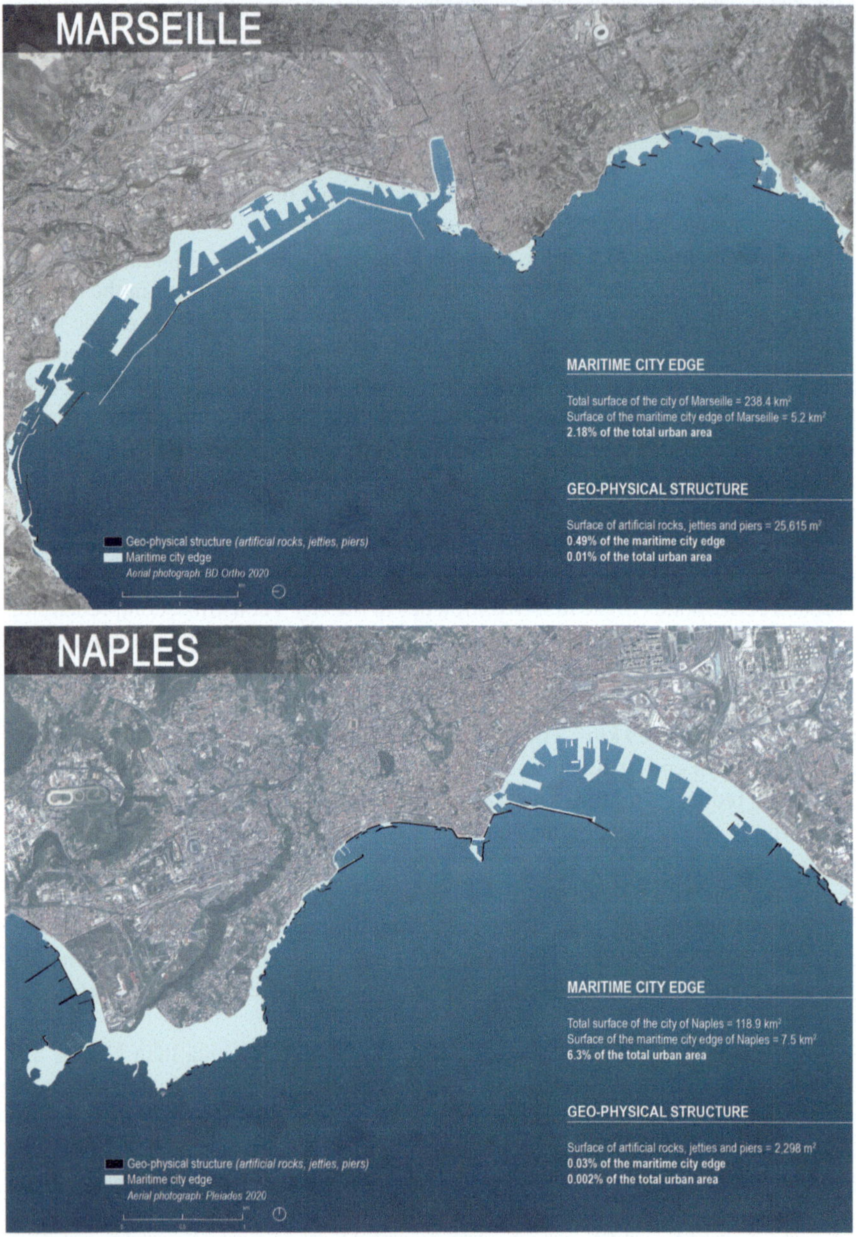

Fig. 13.1 Maps of the geophysical structure within the maritime city edges of Marseille and Naples (elaboration of the author, in collaboration with CNRS ESPACE—UMR 7300)

Fig. 13.2 The Digue du Large in the Marseille port, used for logistic operations and for protection from the sea, is also included in a process for pedestrianization (picture by the author)

allowing it to expand its surface area but also to protect spaces and community uses from environmental risks: think for example of the installation of artificial rocks or breakwaters, but also the use of jetties or floating walkways along the urban coastline (Fig. 13.2).

In this sense, the city of Marseille has a fairly substantial number of artificial reefs whose primary function is precisely to prevent risk from the sea. The percentage of such elements within the local geo-physical structure is 0.49%, or approximately 25,615 m^2. Although this number is statistically low, these elements play an important role in defending the social uses of the urban coastline. In the southern area of the city, for example, the Prado urban beaches are characterised by the presence of extensive arms of artificial rocks that seek to preserve the community function of the place, while also giving a characteristic shape to the area; in the middle portion of the maritime city edge, the rocks present in the area of Malmousque and along the Corniche promenade act as a spontaneous gateway to the sea for local users. On the contrary, in the case of Naples, the geo-physical structure is certainly characterised by the presence of artificial rocks, as well as some jetties, although this functional element quantitatively covers a smaller surface of the coastal interface than its French counterpart: the overall area is only 2,298 m^2. This means that the Neapolitan geo-physical structure measures about 0.03% of the entire maritime city edge and an even smaller 0.002% of the total area of the city. Despite this, it is possible to attribute an important social value to these areas of the city-sea interface. In particular, the coastal area of the Riviera di Chiaia is strongly characterised by the presence of long stretches of artificial rocks and breakwaters parallel to the coast: beyond their role of defence against environmental hazards such as coastal erosion and storm surges, it is a common habit for the Neapolitan coastal community to use these functional

Fig. 13.3 Artificial rocks in the Neapolitan Riviera di Chiaia area, both an environmental defence and a common recreational space (picture by the author)

elements in a social and recreational manner, transforming them into improvised place to sunbath or dive into the sea for bathing; this phenomenon is incredibly common during various periods of the year (Fig. 13.3). Regarding the western area of the city, Bagnoli is famous for its jetties, originally conceived for productive uses related to the area's primary function; however, they are now an integral part of the urban coastal landscape and one of them is also completely dedicated to pedestrian, reinforcing the social-community value of this type of functional asset.

13.3 Natural and Built Environment Along Densely Urbanised Coastal Areas

The natural environment of the maritime city edge of Naples is characterised by a significant amount of vegetated areas: a surface area of 2.15 km^2 characterises its coastal zone, corresponding to almost one third (28.46%) of the total size of the maritime city edge, although this value is equivalent to only 1.9% of the municipal surface area. Vegetated areas include urban parks, private gardens, aligned trees and eventual urban rural areas, but also degraded or abandoned green spaces. The area most closely linked to this aspect of the city-sea interface is undoubtedly Posillipo: the homonymous urban hill is in fact particularly important for the Neapolitan urban coastline because of its naturalistic features, being moreover subject to a specific landscape plan. However, it must be considered that the use of these green areas is limited by the privatisation carried out by residents and economic activities managers, as previously discussed. In any case, it remains the area with the highest concentration of urban greenery in the entire maritime city edge: this clarification is necessary, since other spaces comprehended in this edge of the city-sea interface, although they do not present any particular quantity of urban green, are nevertheless close to vegetated areas. One example is the Riviera di Chiaia and Santa Lucia area, which does not possess a significant amount of open or green spaces, with the exception of the Molosiglio Gardens, nearby the port area: however, immediately beyond the transit road there is the *Villa Comunale* which represents a significant green asset.

In contrast to Naples, Marseille does not show a particularly large area devoted to the natural environment. The data shows that this function occupies 587,521 m^2, or 11.29% of the total size of the maritime city edge.

The reasons include the fact that the maritime city edge tends to be very small, with the exception of a few cases, and the presence of the very large area occupied by the GPMM. The points where urban greenery is most concentrated in the northern area coincide with the Jardin de Corbière, an area that also hosts the only real public access to the sea for bathing in the 16th Arrondissement. Proceeding towards the historical city centre, not far from the Vieux Port, the Jardin du Pharo emerges, a public park accessible to the littoral community where buildings with historical and cultural value are located: however, unlike the Jardin de Corbière, this space does not allow for direct physical contact with the sea. Finally, the Rade Sud offers probably the most important vegetated area of the city, namely the Parc du Prado, which constitutes a public open area equipped for recreation and leisure, as well as for bathing; as in the case of the Riviera di Chiaia, also here it is possible to note that in the immediate proximity there is a large green sports area, whose environmental continuity with the park is interrupted by the transit road (Fig. 13.4).

On the other hand, as far as the built environment of the maritime city edge is concerned, the analysis identifies a value of 14.03% relative to the buildings along the coast of Naples.

The respective 1.06 km^2 of buildings of various types (residential, productive, historical-cultural) represent a significant surface, indicating that a consistent portion of coastal land is densely urbanised. In particular, the areas that show the highest rate of construction are Posillipo and San Giovanni a Teduccio: in the first case, these are mainly residential and accommodation buildings, for a total of 287,806 m^2 corresponding to 27.2% of the total value; in the second case, a large part of the buildings have productive and logistical functions, considering the proximity to the port, and represent more than half of the total constructions present along the maritime city edge. In Marseille, the incidence of the built environment along the maritime city edge is significantly lower in terms of percentage: the 496,460 m^2 of buildings represent 9.54% of the total surface area of the coastal area between the sea and the transit road (Fig. 13.5). The Rade Sud appears most dense in terms of civil urbanisation. In particular, the areas surrounding the *Jardin du Pharo* and Malmousque neighbourhood comprehend 153,563 m^2 of built surface, equal to 31% of the analysed category:

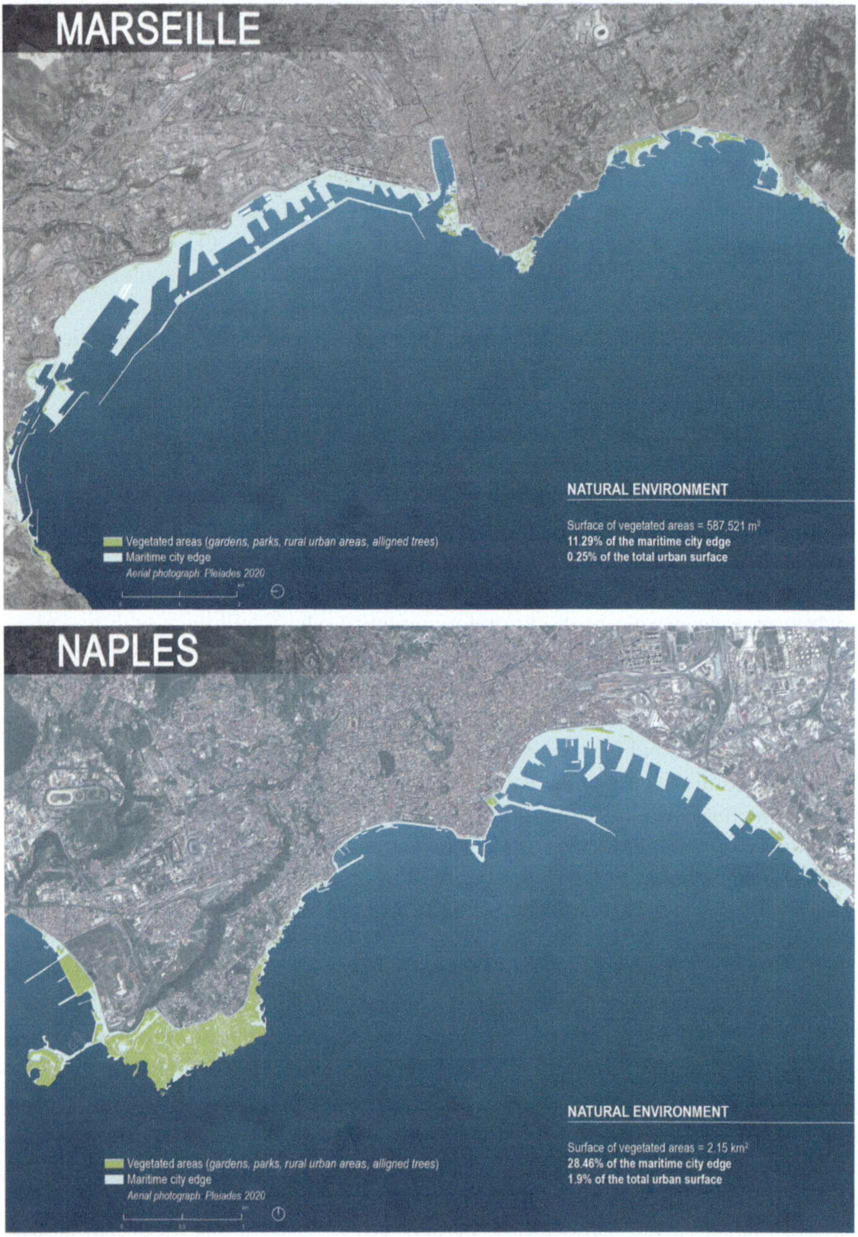

Fig. 13.4 Maps of the natural environment within the maritime city edges of Marseille and Naples (elaboration of the author, in collaboration with CNRS ESPACE—UMR 7300)

these are mainly residential buildings, with some elements with a productive function (Fig. 13.6). Proceeding towards the southern end of the maritime city edge, the Montredon area hosts a further portion of buildings for predominantly residential use, consisting in 95,100 m^2, or 18.4% of the total built environment.

Another element of great relevance within the two city-sea interfaces is undoubtedly the port infrastructure: in both cases, it occupies a substantial portion of the coastal territory, with tourist but mainly commercial and logistical functions, a factor that clearly collides with the development of coastal social and recreational activities and with the free use of the urban coastline. In Naples, the surface area dedicated to the port in its various forms corresponds to approximately 2.26 km^2: a substantial portion of the maritime city edge equal to 30.1% of the total area, mainly located in the easternmost area of the case study. With regard to the French city, this value is even higher: in fact, it is 57.29% of the coastal strip between the transit road and the urban sea: the approximately 3.00 km^2 are mainly concentrated in the central-northern area and correspond to the Grand Port Maritime de Marseille, an infrastructure that is as relevant from an economic point of view as it is limiting in terms of accessibility to the water body for the urban coastal society. A peculiar characteristic of Marseille is the presence of numerous nautical clubs, whose main location is the Vieux Port, in the historical heart of the city, within the Pointe Rouge pole in the Rade Sud, and in the area of L'Estaque, in the 16th Arrondissement: in the latter place they often constitute an insurmountable barrier for the city's users.

A further element that differentiates the French and Italian coastal built environments is the presence in the Marseilles city of large spaces dedicated to the storage of private boats: while in the Italian case study no significant points destined for this function were identified, along the French coast 241,238 m^2 of this kind of constructions, some of them very tall, have been mapped. This represents 4.63% of the total area of the maritime city edge, mostly located along the coast of the 16th Arrondissement (Fig. 13.7).

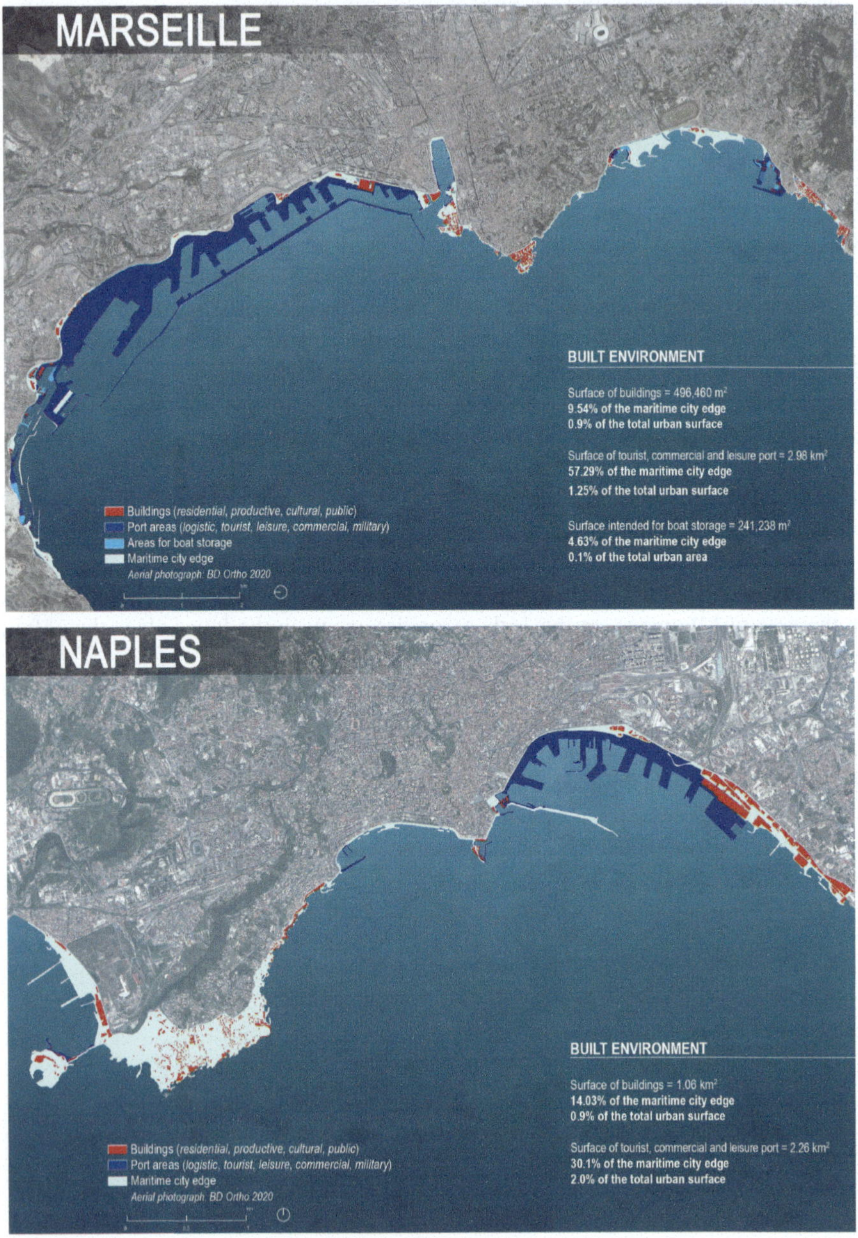

Fig. 13.5 Maps of the built environment within the maritime city edges of Marseille and Naples (elaboration of the author, in collaboration with CNRS ESPACE—UMR 7300)

Fig. 13.6 The Jardin du Pharo, in which the homonymous historic building stands, represents a point of significant urban vegetation within a highly urbanized coastal area in the Rade Sud of Marseille (from Wikimedia Commons, licensed under CC BY-SA 4.0)

Fig. 13.7 Boat storage sites in the Saumaty area, in the Rade Nord of Marseille (picture by the author)

13.4 Main Coastal Recreational Uses and Soft Mobility System

The study of the socio-recreational aspects of the maritime city edge involved the mapping of the main bathing spots in the two cities taken as case studies (Fig. 13.8). Along the Neapolitan coastal interface, it is possible to identify a total area of 142,341 m^2 dedicated to this activity.

This is 1.9% of the total area of the maritime city edge, a value that clearly has little bearing on the vastness of the coastal area of reference. It must be added that the urban beaches in Naples are very diverse, with different degrees of accessibility and water quality: the coast of San Giovanni a Teduccio is forbidden for bathing, due to very high levels of pollution, yet the local community strongly desires to enjoy it; the Riviera di Chiaia has one of the few free Neapolitan beaches, although the provision of bathing-related services is not very efficient, while the available space is quite limited; Posillipo probably offers the widest variety of urban beaches in the city, even if access tends to be private or very expensive, while accessible public areas (such as the Gaiola beach) can only be reached by uneasy routes; finally, the Bagnoli area has urban beaches, but also here contact with the water is forbidden as a consequence of the poor ecological quality of the sea.

In the case of Marseille, the urban beach endowment of the maritime city edge is 148,871 m^2. There is a strong similarity with the Neapolitan case, but this figure reflects a significantly higher surface percentage (2.86% of the total coastal area), although even in this case it is a rather small amount. It should be emphasised, however, that the offer of bathing areas along the French coastline is much more adequate to meet the needs of urban coastal society than its Italian counterpart.

In particular, the Rade Sud appears to be very much oriented towards the recreational use of the urban coasts, thanks to the presence of relevant urban beaches, such as Pointe Rouge and Prado areas: in both cases, the sites are well equipped for bathing and the fruition of the sea resource is completely free of charge; going up towards the central part of the city-sea interface, the Plage des Catalans represents one of the most frequented places by Marseille users, not only for bathing but also for outdoor activities. It should be noted, however, that the central-northern sector is strongly characterised by the port infrastructure: thus, only in the extreme northern part of the maritime city edge it is possible to find additional urban beaches, although in the Rade Nord they are much more limited (Figs. 13.9 and 13.10).

Sports activity and the public spaces designed to host it are a further analysis criterion related to coastal sociality. The digitisation of the two coastal interfaces highlights the fact that in both cases the presence of these elements is relatively low: the share in Naples is 74,715 m^2, equal to 1% of the maritime city edge, compared to the 13,780 m^2 that Marseille devotes to this function (0.26% of the maritime city edge). With regard to these percentages, it must be considered that, although located outside the maritime city edge in the strict sense, the two city-sea interfaces nevertheless offer sports facilities to their users; it must also be said that the Prado area in Marseille constitutes a large multifunctional space, in which open-air sports

Fig. 13.8 Maps of recreational uses and soft mobility in the maritime city edges of Marseille and Naples (elaboration of the author, in collaboration with CNRS ESPACE - UMR 7300)

Fig. 13.9 The small pedestrian promenade in L'Estaque, in the 16th Arrondissement of Marseille, runs alongside the numerous local nautical clubs (picture by the author)

Fig. 13.10 Plage de Corbière in the northern part of the Marseille city-sea interface attracts numerous users, but connections with the city centre are weak (picture by the author)

activity is only one of the possible uses. In the Neapolitan case, on the other hand, it should be noted that most of the outdoor sports facilities present in the maritime city edge are not freely usable by the community as they are included within private nautical clubs (Fig. 13.11).

Fig. 13.11 Sports fields near the Molosiglio gardens: these facilities along the urban coast of Naples are accessible only to members of the reference nautical club (picture by the author)

With regard to the soft mobility offered by the Neapolitan coastal interface, the area dedicated to this function within the maritime city edge is 387,346 m^2, or 5.1% of the total surface area. This percentage includes cycle-pedestrian paths, sidewalks, pedestrianised areas and public squares: in short, this is a parameter that attempts to calculate the spaces in which the pedestrian can move freely according to the current design of the urban amphibious. It should be emphasised, however, that this value does not always reflect direct accessibility to the coast, indeed in most cases pedestrian routes are quite far from the coast, sometimes even in terms of views, as is the case at Posillipo for example. The stretches of coast along which user can easily walk while enjoying visual contact with the sea of the city are mainly located in Bagnoli and along the Riviera di Chiaia: the paths tend to follow the coastline and, despite the presence of some obstacles (especially in the former case), it is generally possible to stroll following the shoreline. In relation to the allocation of parking areas, it must be pointed out that the 102,410 m^2 available along the city-sea interface (equal to 1.36% of the maritime city edge of Naples) guarantee relatively simple movements to the coastal society's users: in fact, almost all of this surface area indicates simple roadside stalls, with the presence of a few paid parking spaces. However, the overall distribution of this function is not optimal, as it is not evenly distributed among all areas of the urban coast, making some places difficult to reach and enjoy, not only in the more peripheral parts of the littoral interface but also in the central areas of the coast (Fig. 13.12).

For the Marseille coast, the surface area dedicated to bicycle and pedestrian mobility is slightly larger than the Neapolitan one: the mapping of the maritime city edge shows 396,350 m^2, representing 7.62% of the total size of the study area. This information is therefore even more favourable in the French case study, since

Fig. 13.12 The majority of the Neapolitan urban beaches are located in the western part of the coast but they are often comprehended in overcrowded private bathing establishment (picture by the author)

Marseille maritime city edge is considerably smaller than its Neapolitan counterpart. To this must be added that almost the entire Rade Sud allows the urban system to be very close to the sea. The Vieux Port and MuCEM areas also offer the same possibility; however, the whole central portion, occupied by the massive Marseilles port, does not allow physical or at least visual permeability towards the coast and the water. The situation does not improve going northwards, in the 16th Arrondissement, as the nautical clubs and the presence of vast areas destined for boat storage make the sea effectively unreachable, both in terms of physical access and in terms of visual contact. Similarly to the Italian situation, the space intended to park two and four-wheeled vehicles covers a very low percentage of the coastal interface, that is equivalent to 1.96% of the maritime city edge. The 102,118 m^2 of parking spaces for cars and motorbikes are, however, discretely distributed, providing a valuable support to local mobility, especially in the Rade Sud.

13.5 Dense Interfaces with Limited Coastal Space for Recreational Use but High Social Potential

The geo-spatial research has provided a clear picture of the functional arrangement of the littoral area which tends to emphasise coastal socio-environmental relations to the greatest degree and to absorb pressures from the urban and maritime spheres, within the city-sea interface theory. The study of the maritime city edges of Marseille and Naples has thus revealed similar characteristics as well as elements of dissonance in these two case studies. As mentioned above, the different overall size of the two city-sea interfaces is linked to their specific spatial conformation but also to the layout of the transit road. Partly, this is due to specific functional reasons, since the French city tends to favour a road infrastructure as close as possible to the sea, while this happens more rarely in Naples; however, considering the morphological structure of the two settlements, it must be considered that the Italian city has a much higher and jagged coast than the French one, which obliges to put a greater distance between the transit road and the shoreline. Hence the different widths of the two coastal interfaces: in Marseille, it tends to widen at a few points, near the port and the limited extrusions of the coast where small settlements arise, such as the village of Malmousque, or in relation to the areas of land reclamation, such as the Parc du Prado; the Neapolitan maritime city edge, on the other hand, assumes a considerable transversal extension, especially in the Posillipo hill, where the highest points of the coast can be found.

However, in spite of the flexibility of the middle component of the city-sea interface, it must be said that Marseille and Naples dedicate a small percentage of the total area to social and recreational functions. In France, this type of use, which is crucial for the purposes of this research, covers just 3.13% of the overall maritime city edge; similarly, in the Italian case, this value reaches 2.88%: these percentages are among the lowest within the functional system of the two urban coasts, an information that must certainly be interpreted in relation to the reference context: in Marseille, for example, there are entire public areas dedicated to open-air recreation and bathing; the Neapolitan coastal interface is less organised, because urban beaches and other socio-recreational facilities are not equally accessible or distributed. Anyway, it should be noted that the Italian case study has a much higher degree of naturalness than the French one: the Neapolitan maritime city edge offers to the local community a considerable amount of urban greenery, in various forms, although not all of it is freely usable; at the same time, in Marseille most of the coastal open spaces can be easily accessed, although the majority of them are located in the Rade Sud.

With regard to these reflections, it is impossible not to refer to the cumber some presence of ports within the case studies. Approximately one third of the Neapolitan maritime city edge is absorbed by this function, while in Marseille the port infrastructure exceeds half of the surface available within the research area. It is clear that both cities are strongly linked to harbour activity, since they are among the major logistical, tourist and commercial poles of the Mediterranean basin; in any case, the percentages found are relevant also because the port constitutes a strong limitation to the visual and functional permeability of the city-sea interface. If in Naples it is still possible for pedestrian to move quite freely around the port terminal even if they don't have to take a ferry (although there aren't particular recreational or social activities), in the Marseille port usability is completely precluded unless the user has a boarding ticket. This opens up interesting scenarios regarding the difficult relationship between the harbour and the urban system that often collide with each other; however, in both Naples and Marseilles, the Port Authority has signed important strategic documents for the community transformation of the coastal areas. In particular, the French case shows a more effective implementation of project operations for the city-sea interface than the Italian case.

In conclusion, it should be noted that the two maritime city edges show a different way of managing space. Marseille is extremely affected by the presence of the port, which significantly limits the functional permeability towards the sea, but denotes the capacity to organise limited spaces in a quite efficient way, despite the geomorphologic or functional limitations, exploiting residual areas in a positive way, even if some critical issues related to conflicts of interest between among its urban coastal society still remain unsolved, like the tensions between ordinary citizens and users of nautical clubs in the northern area of the coast. On the other hand, the coastal interface in Naples has a wider surface but suffers from a discontinuous distribution of public facilities along the shore; moreover, accessibility to the littoral resources is often compromised by unclear governance models, by the privatisation of spaces nearby the urban sea and by the presence of degraded areas and disused sites in the peripheral zone of the maritime city edge, as a result of a tendentially static planning over the last decades. However, the larger size of the Italian maritime city edge may be an advantage in the functional remodelling of the urban amphibious (Table 13.1).

Such reflections suggest that the elaboration of a spatial data model describing the main characters and functionalities of the city-sea interface may actually serve as a basis for planning reasoning: understanding the spatial dynamics underlying coastal functioning may in fact facilitate the identification of sensitive points along the urban littoral and highlight the possible presence of conflict zones. This type of approach can certainly be extended and adapted to different contexts, providing an analytical database that is propaedeutic to a coastal-maritime spatial design responsibly oriented

Table 13.1 Comparison between the amount of elements underlying the coastal interfaces of Marseille and Naples (elaboration of the author)

		Marseille		Naples	
Layer	Elements	Surface	Percentage	Surface	Percentage
Geo-physical structure	Artificial rocks, jetties, piers	25,615 m^2	0.49	2,298 m^2	0.03
Natural environment	Vegetated areas	587,521 m^2	11.29	2.15 km^2	28.46
Built environment	Buildings	496,460 m^2	9.54	1.06 km^2	14.03
	Port	2,98 km^2	57.29	2.26 km^2	30.1
	Boat storage	241,238 m^2	4.64	0 km^2	0.0
Recreational uses	Bathing areas	148,871 m^2	2.86	142,341 m^2	1.9
	Outdoor sport areas	13,780 m^2	0.26	74,715 m^2	1.0
Soft mobility and accessibility	Pedestrian spaces	396,350 m^2	7.62	387,346 m^2	5.1
	Parking	102,118 m^2	1.96	102,410 m^2	1.36
Other uses	Roads, other spaces	210,300 m^2	4.0	1.35 km^2	18.0
Maritime city edge		5.2 km^2		7.5 km^2	

towards the social demand, within the theoretical innovation of the city-sea interface as a link between land and sea planning.

References

Klain SC, Chan KMA (2012) Navigating coastal values: Participatory mapping of ecosystem services for spatial planning. Ecological Economics 12: 104–113

Loomis DK, Paterson SK (2014) The human dimensions of coastal ecosystem services: Managing for social values. Ecological Indicators 44: 6–10

Ranganathan J (2007) Restoring Nature's Capital: An Action Agenda to Sustain Ecosystem Services. World Resources Institute, Washington, D.C.

Chapter 14
Psycho-Social Approach Applied to the Urban Amphibious in Marseille and Naples

Abstract The application of the psycho-social approach to the city-sea interface of Marseille and Naples is hereby illustrated: specifically, the sociological survey focused on four similar urban areas in the two cities, namely productive and port areas, residential green areas, dense urban areas and urban beaches. The objective is to analyse how the various members of the urban coastal society enjoy the urban sea and coastline in relation to different characteristics of the city. Therefore, social perceptions, possible scenarios and awareness of the ecological and environmental dynamics underlying the sustainable development of the urban amphibious have been investigated.

Keywords Psycho-social approach · Social perception · Coastal community · Environmental awareness · Scenario · Recreational uses · Quantitative survey

14.1 Questionnaire Dissemination and Types of Study Areas

The study of the dynamics underlying the social demand of the coastal communities of Marseille and Naples involved the psycho-social approach, according to the methodological framework described above. The proposed questionnaire model was therefore disseminated. In order to facilitate the comparison between the different urban areas of the two city-sea interfaces, four types of study areas were identified, based on the similar morphological-functional characteristics existing between the two Euro-Mediterranean cities (Fig. 14.1).

The productive and port areas include the 16th Arrondissement of Marseille and, for Naples, the coasts of Bagnoli and San Giovanni a Teduccio. The 16th Arrondissement, specifically the area of L'Estaque, a former fishermen village, became over time a productive district and was then partially incorporated by the northern part of the GPMM: today it hosts a disused quarry and one of the city's main bathing spots, as well as a large number of nautical clubs that severely limit accessibility to the sea. The neighbourhoods of Bagnoli and San Giovanni a Teduccio are peculiar sites

I. Pistone, *Urban Coasts in Socio-ecological Transition*,
https://doi.org/10.1007/978-3-031-70783-4_14

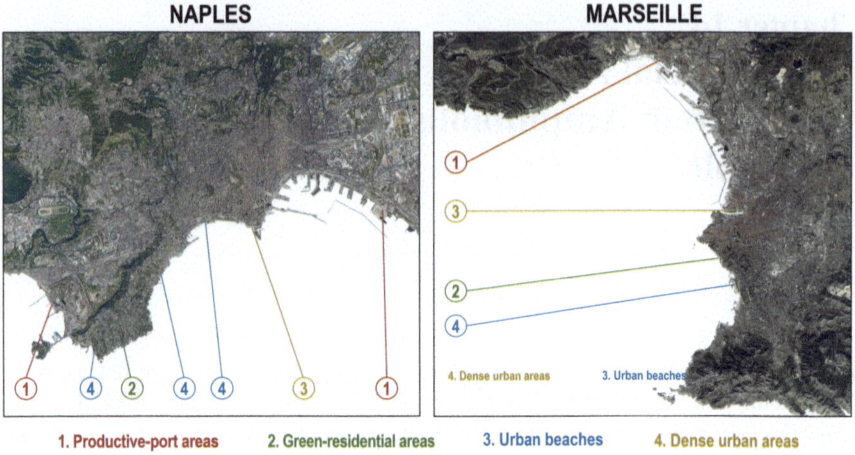

Fig. 14.1 Areas of interest for the survey dissemination in Naples and Marseille for the psycho-social approach (elaboration of the author)

along the Neapolitan coastal interface because local communities express a strong desire to enjoy the coast and the sea, yet the ecological quality of the coastline is deeply damaged by past productive activities that have defined the identity of these places.

Residential green areas include the Corniche area in Marseille and the coastal area of the Posillipo district in Naples. The Corniche, between the area of Malmousque and the Prado bathing area, is a long scenic promenade along which users can enjoy direct and visual contact with the urban sea, passing through a predominantly residential area inhabited by an upper-middle class. Similarly, the Posillipo area constitutes one of the main environmental zones of the urban Neapolitan coastline, as well as a privileged panoramic view: however, the presence of dwellings and properties of wealthier classes and accommodation activities precludes a fair enjoyment of the shore.

As dense urban areas, we consider the historic core of Marseille, specifically the Vieux Port area and the public area of the MuCEM, and the Neapolitan Riviera di Chiaia. The Vieux Port area includes the old harbour of the Marseille city, now transformed into a large pedestrian area in visual contact with the sea, while the MuCEM area is a real space reclaimed by the community, thanks to an agreement with the Port Authority: today it hosts a large square along the shoreline and various cultural functions. The Riviera di Chiaia, on the other hand, is a very popular area along the Neapolitan city-sea interface: here, the residential function is flanked by the tourist and recreational ones, while a wide promenade connects some important elements of the city's historical-cultural heritage; however, bathing is only allowed in some specific points of the zone.

Finally, urban beaches indicate the Prado bathing area for the French case study and some circumscribed areas of the Neapolitan coast, namely the Rotonda Diaz

Table 14.1 Summary scheme of the disseminated questionnaires (elaboration of the author)

Naples									
Typology	Place	Users	Gender			Age			
			M	F	ND	X < 20	20 < X < 40	40 < X < 60X	X > 60
1. Productive and port areas	San Giovanni a Teduccio	32	16	16	0	8	8	8	8
	Bagnoli	32	16	16	0	8	8	8	8
2. Green and residential urban areas	Posillipo	32	16	16	0	8	8	8	8
3. Urban beach areas	Rotonda Diaz, Bagni Regina Elena, Gaiola	32	16	16	0	8	8	8	8
4. Dense urban areas	Riviera di Chiaia	32	16	16	0	8	8	8	8
Marseille									
Typology	Place	Users	Gender			Age			
			M	F	ND	X < 20	20 < X < 40	40 < X < 60X	X > 60
1. Productive and port areas	16th Arrondissement	16	8	8	0	2	4	4	8
2. Green and residential urban areas	La Corniche (between Malmousque and Prado)	36	14	20	2	1	18	9	8
3. Urban beach areas	Prado	16	9	7	0	3	7	4	2
4. Dense urban areas	Vieux-Port, MUCEM	37	19	18	0	6	20	6	5

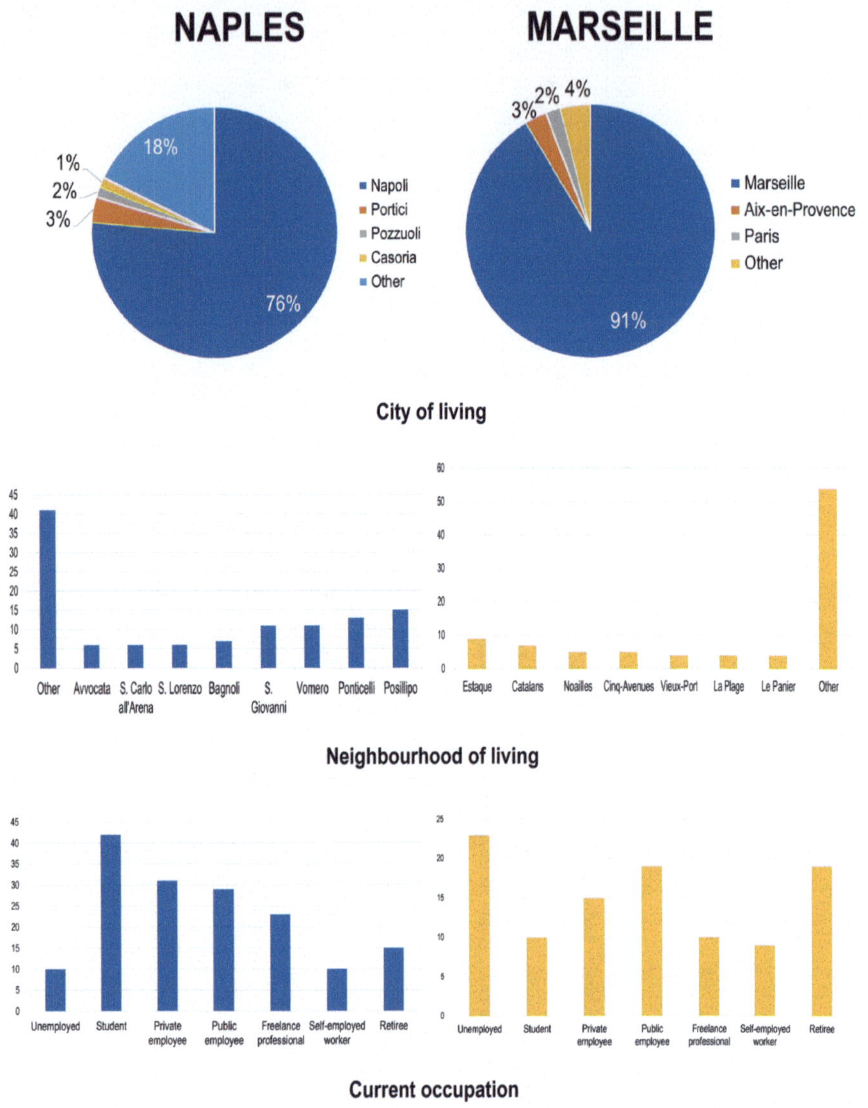

Fig. 14.2 Composition of social samples in Naples and Marseille (elaboration of the author)

beach, the Bagno Elena bathing establishment and the Gaiola beach. As for the Prado, it is a vast zone with a recreational function, almost entirely realised through land reclamation: there are numerous bathing spots but also a large green area dedicated to outdoor sports activities. The bathing spots chosen for the dissemination of questionnaires in Naples are located in different areas of the city-sea interface and present different degrees of accessibility: the beach of the Rotonda Diaz has a public nature although it is generally crowded in summer; the beach of Bagno Elena is located in Posillipo and its access is subject to the payment of a ticket; the beach of Gaiola is comprehended within the homonymous protected park, it is freely accessible by users although it is regulated by strict rules and it is not easy to reach it due to the roughness of the access road.

The questionnaire involved 265 users, 160 in Naples and 105 in Marseille, distributed in the different study areas described above (Table 14.1). The dissemination phase took place during the spring 2022: in particular, the questionnaires were disseminated during the daytime, between 8:30 and 14:30, in order to provide a more precise comparison conditions.[1] The composition of the social samples followed the criteria of the representativeness concept, as explained in the previous chapters (Fig. 14.2). The information found allows for an in-depth analysis of the needs of an urban coastal society sample within the two European case studies; moreover, it can serve as a basis for developing project guidelines for coastal planning that is more aware of the needs of the littoral community and its specific social demand.

14.2 Accessibility of Urban Blue Spaces

Access to the city's blue spaces, understood as the ability to utilise and benefit from the available marine resources or areas of the city-sea interface, is fundamental to the well-being and sustainable growth of urban coastal society. The physical access to coastal spaces for diverse uses, including social and recreational activities, requires careful management with regard to user perceptions: it is therefore an element within the planning of the city-sea interface that is strongly influenced by the concept of socio-environmental justice. In this regard, according to the questionnaire scheme disseminated along the city-sea interface of Marseille and Naples, interviewees from the two cities were asked how important physical contact with the water with regard to activities such as bathing and recreational fishing. The results are quite similar for both case studies, as most users rated contact with the sea as important (46% of the Marseille users and 42% of the Naples users). This value also remains solid for the 'very much' option, with 44% of the answers on the French side and 25% on the Italian side (Fig. 14.3). If these values indeed appear predictable for two cities so closely linked to the sea, a comparison with the frequency of visits for recreational purposes is interesting.

In Naples, there is a discrepancy between the strong interest shown in coming into contact with the sea resource and the percentage of visits to the coast: 39% of the users states to go to the city's coastal area never or rarely (no more than twice a month), while 33% of the analysed sample spends time there about once a week; only 8% of the coastal society declares to visit the urban coast for recreational purposes

[1] Academic collaboration with CNRS ESPACE—UMR 7300 and Aix-Marseille Université enabled the processing of the survey data collected. Moreover, the dissemination of the questionnaires in Marseille has been developed thanks to the field work of the above-mentioned French research institutions.

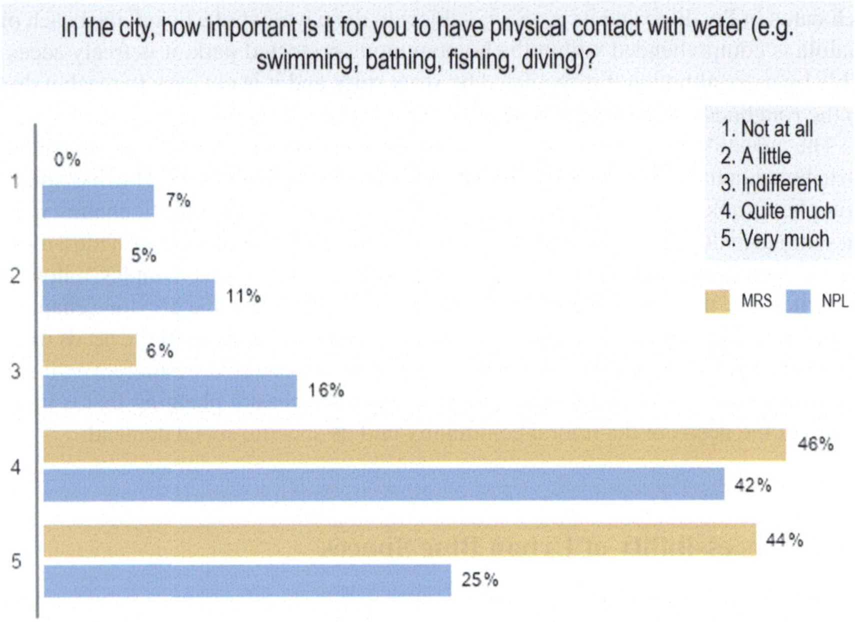

Fig. 14.3 The work highlights a strong enthusiasm in the Marseille and Neapolitan coastal society regarding direct contact with the urban sea (elaboration of the author in collaboration with CNRS ESPACE—UMR 7300)

often (at least three times a week). On the other hand, the habits of the citizens of Marseille show a direct proportionality relationship between the desire to visit the coast and the actual frequency of visits: only 7% of the interviewees admits to spend very little or any time at all along the city-sea interface for leisure and social purposes; more consistent are the values of users who frequent the coast about once a week (27%) or about twice a week (26%); finally, the most consistent figure reflects the 41% of the users involved in the research who admits to visit the coast of Marseille at least three times a week (Fig. 14.4).

The reasons behind the missed satisfaction of the need for contact with the sea can be multiple, related to functional and logistical features. The proposed questionnaire investigates the connection between this criticality and the issue of mobility, both light and heavy, and the constraints to fruition that the coastal interfaces of Naples and Marseilles demonstrate. From the point of view of soft mobility, the two interfaces differ greatly in terms of offered services. More than half of the Neapolitan users has difficulty in reaching the urban coast by foot, while only 14% of the interviewees considers it easy or very easy; as for cycling, 29% finds that going to the shore in this way is hard, while only 11% of users defines it as very easy. In Marseille, the situation is diametrically opposed, as citizens consider that reaching the coast by foot is incredibly easy (38% defines this action as very easy, compared with the lowest recorded value of 12% who defines it as very hard). When it comes to the possibility

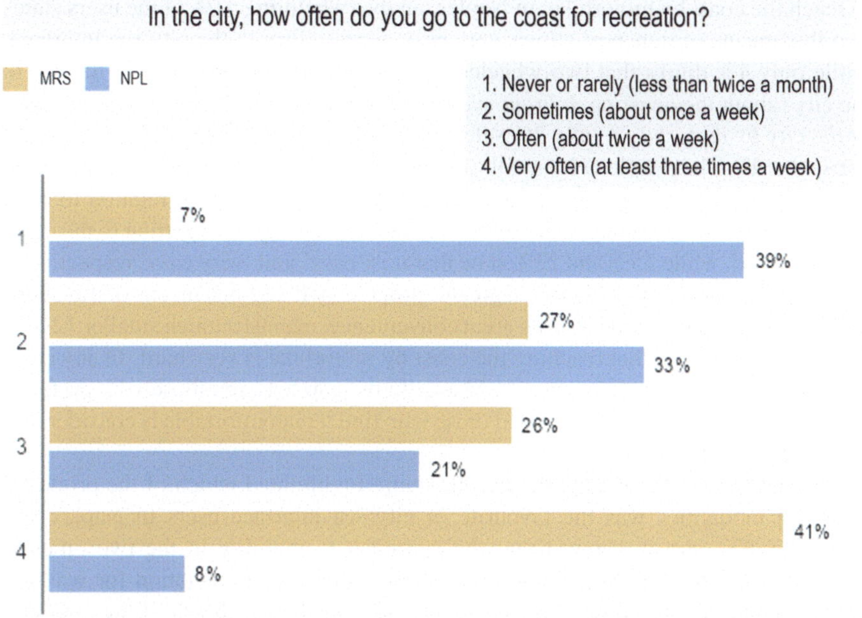

Fig. 14.4 The graphs show a discrepancy between the frequency of visits to the coast by Naples users and Marseille users: in particular, the answers of the former are less consistent with the data collected regarding the importance attributed to physical contact with the city water (elaboration of the author in collaboration with CNRS ESPACE—UMR 7300)

of approaching the sea by bicycle, similar data are found: one third of citizens say that it is easy compared to 16% who says that it is difficult.

Getting around by public transport such as buses is clearly easier for the users of both city-sea interfaces, who share fairly similar opinions: the responses of Marseille interviewees who think that reaching the sea in this way has a difficulty degree within the norm amount to 28%, while 54% of the sample considers easy or very easy to access the coast by bus. Similarly, in Naples, the most registered value is of difficulty is 'within the norm', with 39% of the answers given, showing progressively lower figures in the range of other possible answers. A similar situation is recorded with regard to reach the urban coastline by metro in the two cities: in both Marseille and Naples, the highest number of responses falls on the intermediate difficulty value (respectively 30% and 31% of the users defined the complexity of using this means of public transport as within the norm); however, Naples also denotes a high percentage of negative responses, since 30% of the users opted for the 'very hard' option provided by the questionnaire.

Reaching the coast by car is incredibly easy for Neapolitan users: 42% of the interviewees chose the 'easy' option and an additional 33% opted for 'very easy', in contrast to a not very significant 2% of the sample who consider this means of transport to be very uncomfortable. A similar situation can be found when it comes to reach the coast by motorbike: in Naples, an overwhelming 51% of the users states that this means of transport allows very easy accessibility to the city-sea interface, while only 4% thinks that two-wheeled vehicles are inadequate to go to the sea in the city (about the question *3. From where you live, how easy is it to reach the shore in the city by motorbike?*, only 2% selected the option 'very hard' and a further 2% chose 'hard'). The people of Marseille, on the other hand, have a different conception of going to the coast by private means, on two or four wheels. In relation to cars, users show a certain homogeneity in their answers: 22% says that getting to the coast by car is hard, while 25% and 23% state that it is 'easy' and 'very easy' respectively. With regard to motorbike, there is a greater propensity to use this means of transport, as 32% of the users emphasises its great convenience, against a much smaller 12% of the sample who says that reaching the coast by motorbike is very hard. In any case, compared to the Neapolitan case, in Marseille the gap between those who prefer to reach the shore by private means and those who find it uncomfortable is considerably less pronounced.

On the basis of these data, the questionnaire highlighted which of the proposed transport modalities was the favourite of city-sea interface users in Naples and Marseille. The survey showed that the approach to mobility in the two cities is very different. In fact, Neapolitan users show a discrete appreciation for walking (21%), while the preference for cycling is much weaker, with 8% of the answers, but also for public transportation such as buses and the metro (respectively, 9% and 6% of the studied sample). The most popular means of transport is undoubtedly the car, with 38% of the responses in favour, while moving around by motorbike received only 16% of the preferences, although it was perceived as the easiest way to reach the coast. On the Marseille side, the trend is reversed compared to the Naples case: users show much less interest in using cars (15%) and motorbikes (5%); the use of public buses reaches 14% of the preferences, similarly to subways, which account for 15%; undoubtedly, Marseille users prefer to reach the urban coast by foot, according to 37% of the responses, while going to the shore by bicycle totals 22% of the preferences (Fig. 14.5).

In addition to transport preferences towards the urban coastline, the constraints limiting the use of this resource within the city-sea interface of Naples and Marseille were also explored. From the results of the disseminated questionnaire, it emerges that physical barriers seem to be the limitation that users in both cities perceive as least impacting on the enjoyment of the urban sea. In Naples, 28% of the citizens in fact believes that physical barriers along the urban shoreline have little influence on general accessibility, while 15% states that they are not an issue at all; however, it should be emphasised that a substantial 27% considers them to be very limiting to the fruition of the coast. On the other hand, the Marseille inhabitants involved in the survey show an even more low perception of physical constraints to the city's water: 27% and 39% of the users consider them to have respectively little or any impact on the use of the urban coastal resource; anyway, 15% perceives physical barriers as a strong obstacle to maritime recreation. In relation to the conditions of pollution along the coast of the two cities, in Naples, 43% of the interviewees perceives them as having a strong impact on maritime recreation (26% even consider

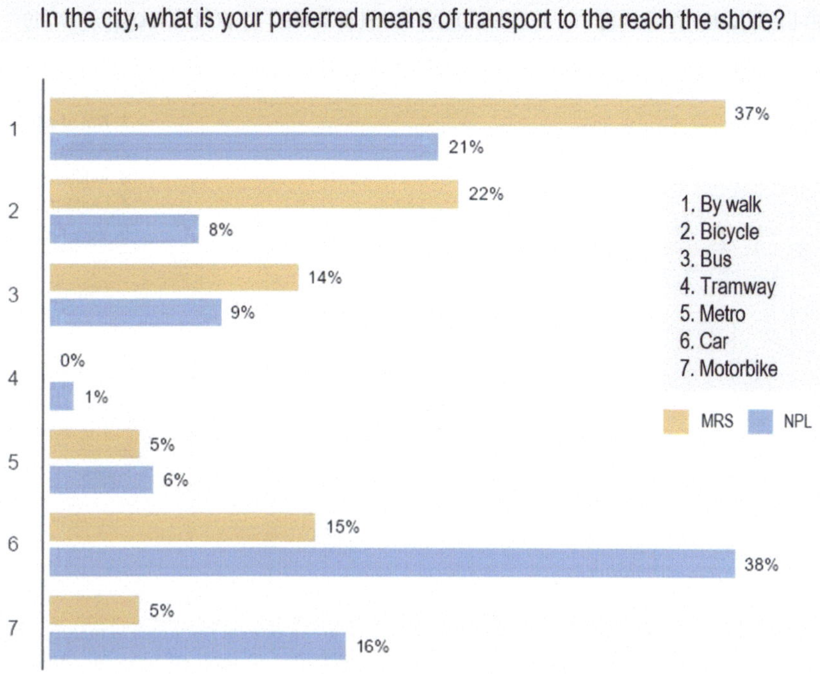

Fig. 14.5 Analysis of preferences regarding transport to the coast, which highlights a clear difference between the Marseille and Naples samples (elaboration of the author in collaboration with CNRS ESPACE—UMR 7300)

them to be incredibly impactful); similarly, in Marseille, 45% of the users perceives a strong constraint in relation to environmental pollution (Fig. 14.6). The results for this type of constraint on the use of the city-sea interface appear firstly much more incisive than the perception of physical coastal barriers, and secondly quite similar between the littoral interfaces of the two case studies. Finally, the perception of eventual legal constraints to maritime-coastal accessibility is analysed. Again, the results that emerged along the Neapolitan city-sea interface tend to highlight a marked perception of the limitation imposed by this kind of constraint: 34% considers it to be quite much limiting, while 17% considers it to be very much limiting for coastal fruition; in any case, the middle option indicating a general indifference towards legal constraints is a relevant figure because it groups together 21% of the total answers. On the other hand, the opinions of the Marseille coastal society are more evenly distributed over the range of possible options regarding the perceptions of coastal legal constraints, while the incidence of users who consider this type of limitation on littoral accessibility really impactful is very low, with only 6% of the answers.

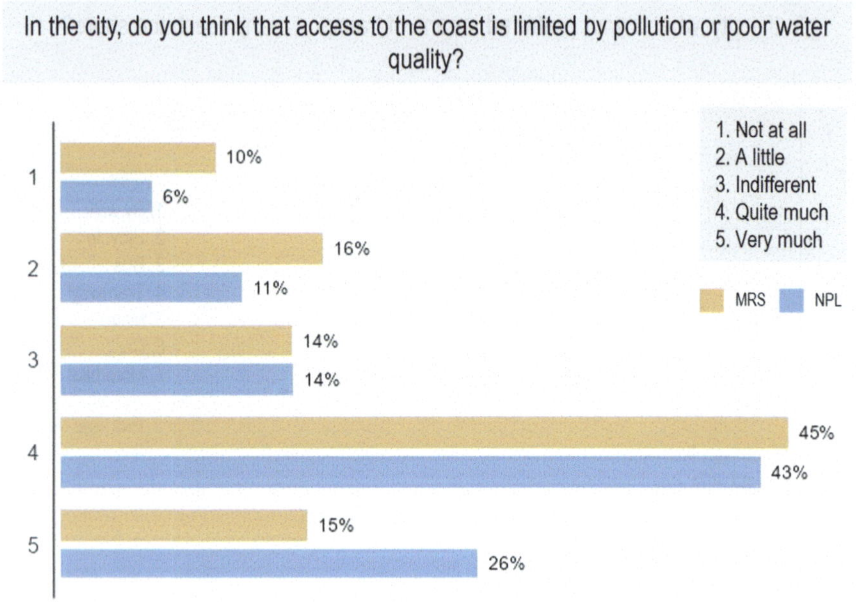

Fig. 14.6 Users' perception of constraints on the use of urban coastal resources is mainly linked to poor water quality and environmental pollution, rather than to legal or physical limitations (elaboration of the author in collaboration with CNRS ESPACE—UMR 7300)

14.3 Land and Sea Uses in the City-Sea Interface

The study of the functionality of the city-sea interfaces of Naples and Marseille is articulated through an analysis of how the urban coastal society perceives the endowment of public areas near the city's coastline. Concerning the presence of pedestrian areas or coastal promenades, users in both cities seem quite satisfied with the current situation: 32% of the interviewed Neapolitans declares that the shore of their city is sufficiently equipped with this kind of public spaces, while an important 29% also states that there are even a conspicuous number of them; in Marseille, 35% of users asserts that there is an adequate number of coastal pedestrian areas, although it is necessary to point out that 23% of the interviewees considers the amount of such areas along the urban littoral to be low, while 24% believes that the city's coastline does not meet the minimum functional requirements at all.

Regarding the presence of coastal green areas such as parks or urban gardens, the results from the two case studies generally agree on the lack of this type of socio-recreational element along the coastal interface. Both samples of urban coastal society think that parks and gardens are available to a limited extent along the shore, particularly in Naples 41% and in Marseille 36%; it is also worth noting that 30% of the French users emphasises the absolute insufficiency of this littoral asset. In any case,

29% of the Neapolitan citizens and 23% of the Marseille citizens consider themselves moderately satisfied with the current situation, while a much smaller number of users thinks to an above-average presence of green areas. A similar situation is found with regard to the presence of spaces for outdoor sports activities. 39% of the Naples users defines such public spaces as quite lacking, similar to 34% of the Marseille survey sample, while 36% of the citizens of the Italian case and 29% of the French ones consider them to be totally insufficient. It should be highlighted, however, that a certain percentage of citizens involved in the analysis considers the sports areas along the urban shoreline to be at least sufficient, especially in Marseille where 27% of the answers in this sense were recorded; in Naples, a more lukewarm 19% emerged instead.

Urban beaches play an important role in this work, yet the perception of their presence is not so obvious within the coastal communities of the two case studies. Users in Marseille consider urban beaches to be below the minimum standard necessary to meet local socio-recreational needs, with 37% of the answers, while 14% of the users considers them to be totally insufficient. 22% of the French sample thinks that the current number of urban beaches is sufficient, while 17% considers them more than sufficient. On the Italian side, the situation is roughly the same: although 26% considers that the city-sea interface in Naples offers a sufficient amount of beaches, only 8% considers this kind of element more than sufficient, while 41% thinks that it is insufficient (Fig. 14.7).

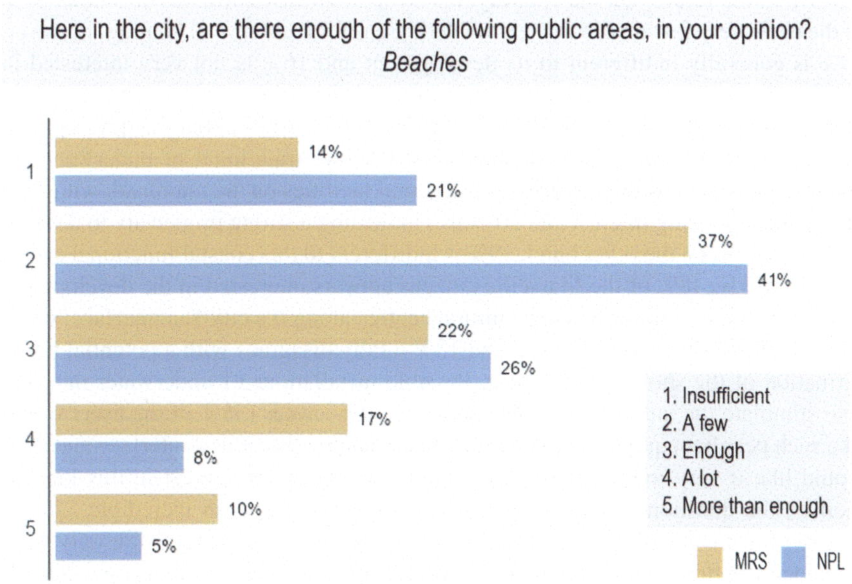

Fig. 14.7 Both city-sea interfaces examined in the book are lacking of beaches in the eyes of the users to whom the questionnaire was submitted (elaboration of the author in collaboration with CNRS ESPACE—UMR 7300)

The questionnaire therefore investigated the needs of users in Naples and Marseille with regard to facilities along the urban coastline and their potential improvement. Marseille users showed a particular aversion to the development of receptive activities such as restaurants or bars: 47% of the interviewees would not like them to be developed at all, while only 6% would like them to be greatly expanded. In Naples, opinions were more homogeneous: although not the highest value recorded, a good 20% would like accommodation activities to strongly develop; overall, 52% of the users would prefer little or no development of bars or restaurants along the coast, while 22% is indifferent to such littoral change. The possible future installation of facilities to rent nautical equipment (such as kayaks or diving harnesses) is much more appreciated, particularly in Marseille, where about half of the users surveyed would like such a development for the urban amphibious, while 26% shows indifference; only 9% confesses to be uninterested. In Naples, however, there is a warm welcome for a possible increase in such recreational use of the city-sea interface, with 34% of the users declaring very favourable opinions, compared with 24% of the interviewees who is indifferent to the implementation of this activity; however, just over a fifth of the users (21%) shows low interest.

According to the data collected, the needs of citizens in Marseille and Naples diverge sharply in relation to the development of tourist docks and private harbours: the former are particularly opposed to this function (45% of the users would not like to see any further increase in the areas for recreational port activities), although 14% of the interviewees is indifferent to the issue and 18% would be interested in seeing it develop in the future. On the contrary, the Neapolitan case shows a substantial 35% of the users very favourable to the growth of areas for recreational boating; however, 29% is generally indifferent to its development and 16% is not very interested in it (Figs. 14.8 and 14.9). The information provided by the questionnaire shows an overall alignment between the desires of users in both cities about the development of spaces for cultural centres or museums, whether traditional or underwater. In Naples, there is considerable interest in cultural facilities on the mainland, with 46% of the users showing interest and 20% also indicating a strong propensity to develop such activities; on the other hand, 19% is indifferent to this coastal functional evolution. Similarly, 48% of the Marseille citizens appears interested in the development or improvement of museums and cultural centres along the city-sea interface, while a smaller 20% shows indifference and 13% totally disagrees with a potential transformation of the shore in this sense. Possible installations of underwater museum also stimulate the interest of the Neapolitans interviewees (35% of the users would like such peculiar equipment to be implemented along the littoral interface and 31% would like it very much), while 18% shows no particular interest in this kind of facility; the questionnaires disseminated in Marseille show an incredible 57% of the users in favour of the development of underwater museums, while only 17% remained indifferent to this potential intervention: this figure is interesting, considering that Marseille is one of the very few examples in the world, and the only one in the Mediterranean, to boast an underwater museum currently existing and active in the Anse des Catalans.

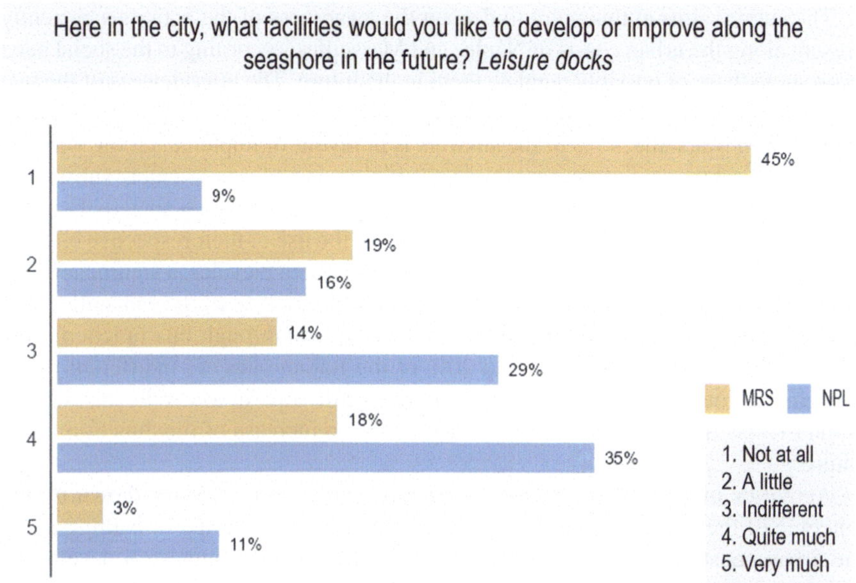

Fig. 14.8 According to the survey, there is a difference between the desires of Marseille and Naples citizens regarding the tourist-port development of the city-sea interface (elaboration of the author in collaboration with CNRS ESPACE—UMR 7300)

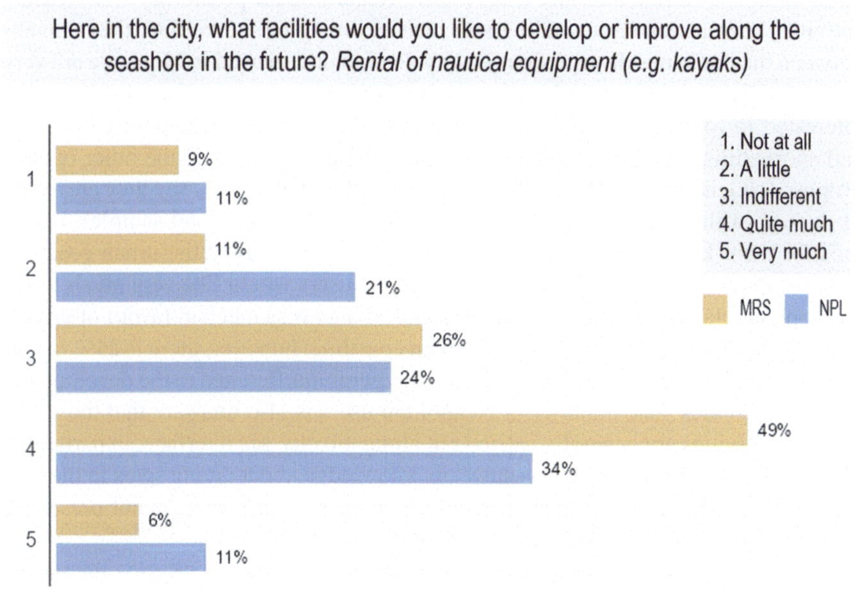

Fig. 14.9 The graphs highlight that the hypothetical increase in facilities for the rental of nautical equipment is well considered in both cities (elaboration of the author in collaboration with CNRS ESPACE—UMR 7300)

The survey aims at this point to deepen the awareness of the activities currently present along the urban coasts of Naples and Marseilles, according to the social need to promote them or possibly regulate them in the future. The interviewees of the two city-sea interfaces show a strong desire for having the possibility to walk along the shore: in Marseille, 47% of the answers is in favour of implementation, while in Naples 36% of the users expresses the same opinion, in addition to a more substantial 42% who states to be extremely inclined to the promotion of such an activity in the future. Both samples analysed show awareness of the use of their respective city-sea interfaces in this sense, however, with regard to the use of bicycles, a significant 21% of the Neapolitan users and a milder but still significant 12% of the Marseille users affirms to be unaware of the use of the urban coastline through this function: only 18% of the French interviewees and 20% of the Italian ones are indifferent to the implementation of bicycle use along the coast, while 36% in Marseille and 41% in Naples respectively would like very much to see an increase of this function in the future.

Activities in the water attract great interest in the two cities, as shown by the outcome of the analysis. Concerning the possibility of swimming or bathing along the urban coastline, 47% of the users in Marseille affirmed interest in developing this function of the urban sea to a greater extent, compared with 53% in Naples, which shows an even greater interest, while much lower are the values related to indifference or aversion to this practice; in both cities, however, according to about a third of the two social samples, this recreational action is not allowed or does not take place along the coastal interface (Fig. 14.10). Regarding the possibility of increasing activities such as rowing or sailing and such as diving or snorkelling, the demands between the two cities diverge significantly. People interviewed in Marseille are very keen on the development of these functions (56% of the users defines itself as very interested in rowing and 46% says it would like to see the development of diving and snorkelling activities, significantly outstripping the values of the other options proposed). Generally, the Marseille users involved in the analysis say they are aware of the use of the urban sea for boating activities (80% of the total sample), just as most of them know that it is possible to snorkel or dive along the urban coastline (62% of the total sample). In Naples, 37% of the users would like very much to see the implementation of sailing and rowing and 31% is very much in favour of greater future growth of snorkelling along the urban coastline: however, there is 24% for the former and 27% for the latter who express a general indifference to the development of these activities. The majority of Neapolitan users is also unaware that these two types of activities are generally carried out along the city-sea interface: in particular, 80% of the interviewees does not show any knowledge about diving spots in the city, while 57% of the users believes that sailing or rowing functions are not present in the range of actions offered by the Neapolitan littoral.

Generally speaking, fishing finds similar opinions in the two city-sea interfaces: in both case studies, about one third of the interviewees were indifferent to the future development of this practice along the urban littoral; 17% of the Italian users and 18% of the French users would instead like fishing in the city to be encouraged, while 16% of the Neapolitan sample and 19% of the Marseille sample would like

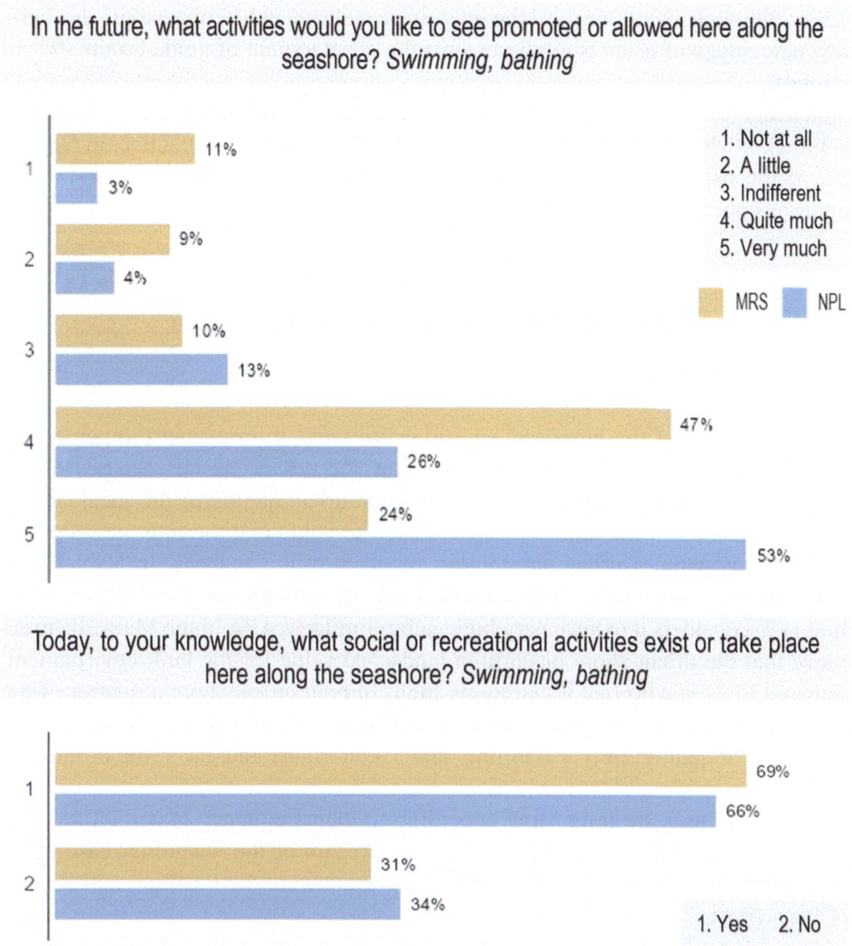

Fig. 14.10 Study of the demand of coastal society regarding future socio-recreational activities specifically linked to contact with water and the relative degree of awareness of existing uses (elaboration of the author in collaboration with CNRS ESPACE—UMR 7300)

this activity not to be developed further in the future. However, while almost all of the Marseille users claim to know that fishing is legally allowed or at least practised within the city-sea interface, 41% of the Neapolitans states to be unaware of this practice. The survey also showed that the social sample studied for the Italian coast has a very strong interest in the implementation of activities related to guided and cultural tours along the urban shoreline: 33% of the users considers itself interested in such a development and 29% denotes an extremely high interest; 17% appears to be simply indifferent to such a coastal development while only 5% is totally against socio-cultural initiatives of this kind. For Marseille, the situation is more or less the same, as 42% of the users shows a propensity for the future development of this activity, although about one third the interviewees shows indifference to it; also here, a low percentage of users is totally against the development of guided tours (9% of the total).

On average, in both cities, it appears that the majority of the interviewed citizens has little knowledge about the current presence of such a coastal practice: in Naples, 62% of the users is unaware of it, while the Marseille residents record 43% of responses coherent with this opinion.

14.4 Perception and Future Scenarios of the Urban Coastline

The investigation of the proposed questionnaire seeks to bring out the perception of coastal space and landscape, studying how the urban coastal societies of the two case studies interpret the environmental and urban quality of the local coastal context. Specifically, the aim is to understand the value that the interviewees give to the various parts of urban amphibious involved in the analysis. In Naples, one third of the users attributes a high value to the built and urbanised area of the shore, while 14% considers it to have very high value; similarly, 41% of the Marseille users believe that the urban shore has a high landscape value for the built environment, compared to 21% who rate it extremely high. In both cities, about a quarter of the analysed social sample considers the urbanised landscape of the city-sea interface to have a neutral quality; only 8% in Italy and 3% in France rate the value of the area very low.

These responses are interesting according to their frequency in certain areas of the Neapolitan and Marseille city-sea interface. Regarding the Neapolitan case study, the interface areas with the highest landscape value were the Posillipo and Riviera di Chiaia coasts.

Neapolitan urban beaches recorded a solid 47% of the responses regarding a high landscape value for the urbanised environment: the reason probably lies in the fact that they are generally located in the most well-kept urban zones of the shore of Naples. The productive-port areas of the Neapolitan coastline record a generally neutral or low landscape value for the built environment: however, the landscape value of the Bagnoli littoral is nevertheless assessed as quite high by 25% of the users. In Marseille, the dense urban areas (namely the Vieux Port and MuCEM zones) and the green-residential area along the Corniche achieved the most positive results. The Marseille urban beach area (in this case, the Prado zone) also registers a relevant 35% of responses describing the surrounding built landscape as very high: it is interesting since it is a large blue-green lung of the urbanised Marseille shore, but it is also located nearby important architectural elements like the Château Borély. In contrast, the productive-port area of L'Estaque receives a lukewarm appreciation from citizens, who rate the urbanised landscape as neutral by 35% (Fig. 14.11).

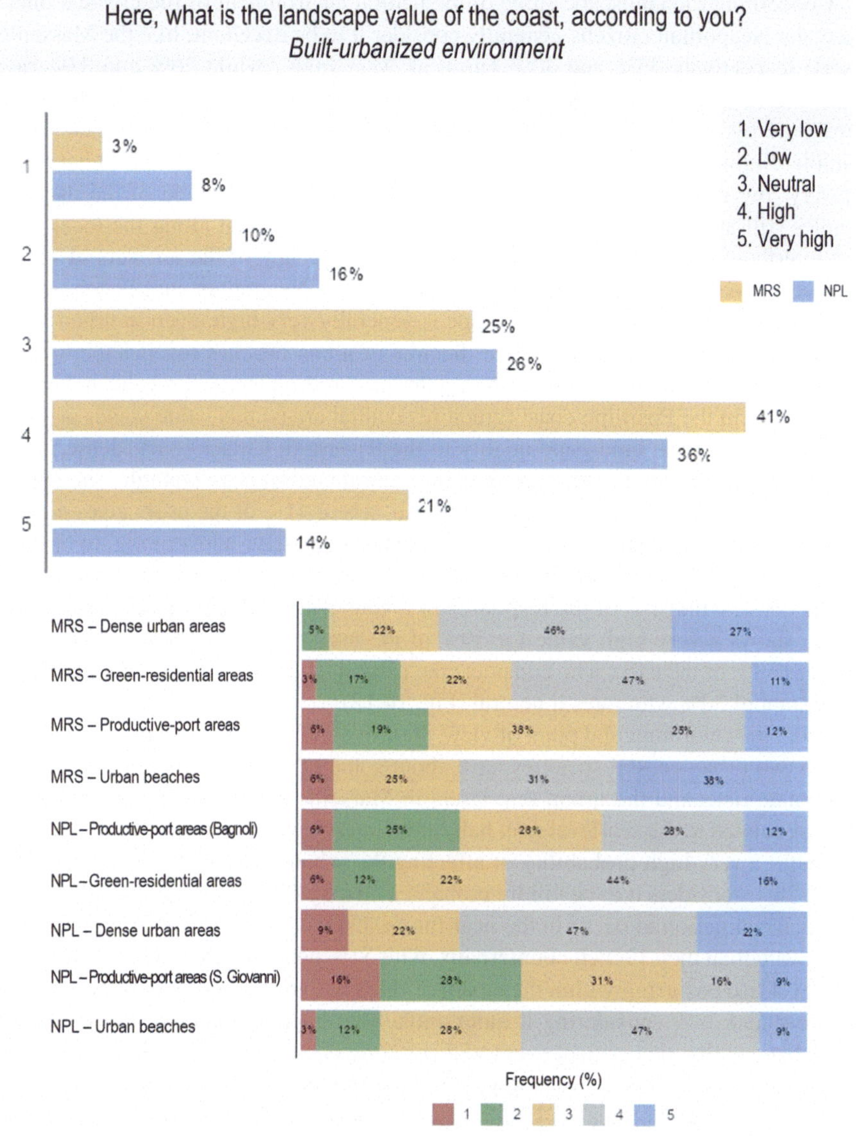

Fig. 14.11 Graph regarding the perception of the landscape quality of the built-urbanized environment. Below, the frequency of the answers given at the different dissemination points (elaboration of the author in collaboration with CNRS ESPACE—UMR 7300)

Concerning the landscape value of the natural environment of the city-sea interface, the Neapolitan citizens generally consider it to be excellent, like the Marseille users: respectively 33% and 40% rate it as 'very high', while 26% and 33% rate it as 'high'. The neutral value of the natural component of the coastal landscape is attributed by only 19% of the Italian social sample and 14% of the French social sample. Significantly fewer users in both cities consider the landscape to have low or very low value: however, it makes sense to point out that 17% of the users in Naples still attributes a low rating to the natural environment along the local city-sea interface. Again, it is useful to analyse the frequency of the answers given by the users involved in the field research. As far as the Neapolitan case is concerned, the environmental and naturalistic value is generally very high, even in urban areas characterised by dense urbanisation: the area of urban beaches reaches the highest values, with 50% of the users considering the surrounding landscape value to be very high, while in the Posillipo coast (green-residential areas) this value stands at 47%; 35% attributes great landscape quality to the Riviera di Chiaia coast (dense urban areas), while in Bagnoli 41% considers the natural landscape to be high. The lowest figure is reached by San Giovanni a Teduccio, where 41% of the users gives neutral quality to the natural landscape and 34% considers it to have a low rating. In France, the Corniche is the area with the highest landscape value for the natural environment of the coast, with 61% of the responses in favour. The landscape of L'Estaque and Prado shows a very high value for 50% of the users, while the dense urban areas register 38% of the users who considers the natural landscape value as high, but also a substantial 30% who rates it neutral (Fig. 14.12).

From the environmental point of view, the work analyses the perception that users of the two urban coastal societies have about some of the main phenomena related to climate risks and the urban sea. Users in Marseille show a strong awareness of the risk related to the sea-level rise: half of the interviewees declares that this critical issue has a very high probability of affecting the urban coast in the next five years, while 28% considers it to be mildly probable and only 6% believes it is hardly likely for such a situation to occur in the near future. On the contrary, in Naples users are less aware than their French counterparts, with 51% believing that the probability of sea-level rise occurring within the next five years has an average level of probability to occur and 29% considering it quite unlikely to happen; only 4% of the Naples users believe that this critical event will certainly occur in the near future. Also for coastal erosion, Marseille users appear to be more alert to the incipient difficulties that their littoral interface will address in the near future: just over a third of the citizens firmly believes in the impact of this risk on the urban coast in the next five years, compared to 17% who perceives this probability on average level and a small 1% who instead considers the eventuality to be completely implausible. Half of the Naples sample believes that the risk of coastal erosion is plausible, but only 4% considers the gradual loss of coastal land a certainty, while 23% of the social sample says to perceive the future threat to be slight (Fig. 14.13).

An increase in the frequency of coastal floods is assessed as probable by half of the Naples social sample, while 34% believes it is not a very plausible eventuality; only 1% assess this risk as a more than sure threat to the Neapolitan coast over the next five

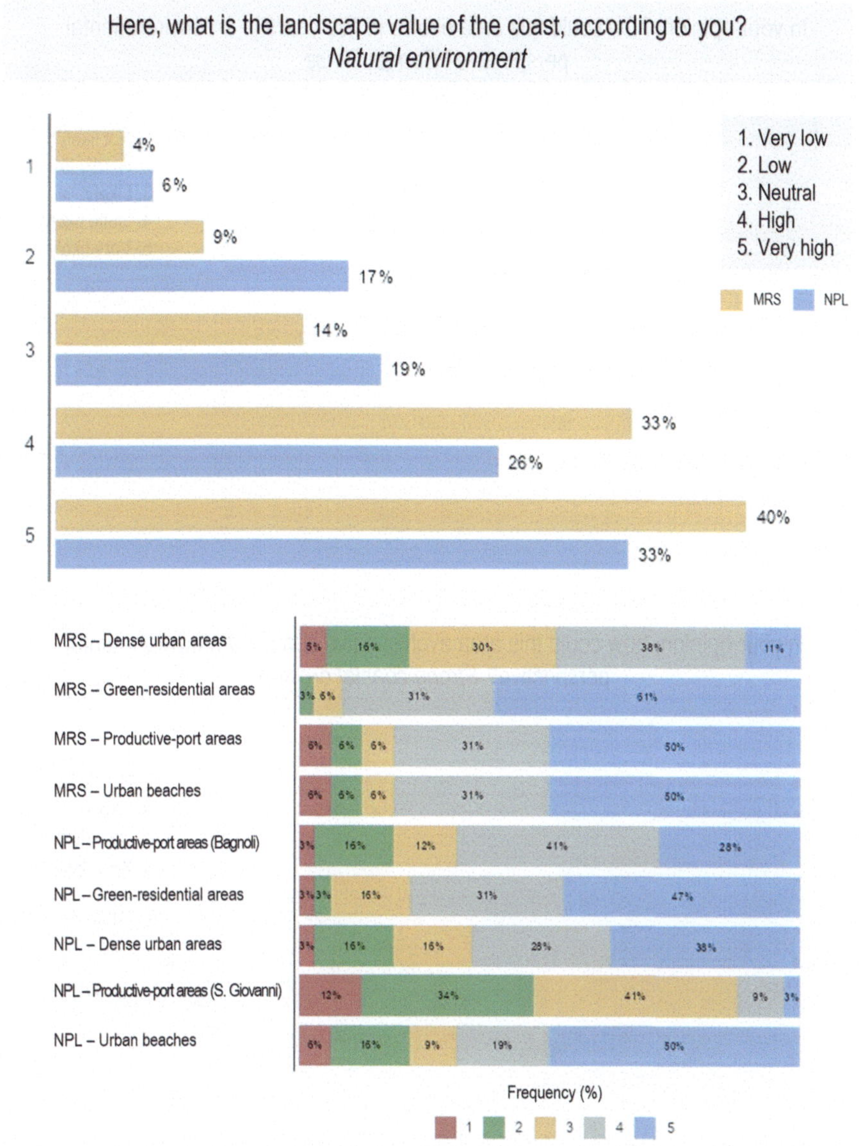

Fig. 14.12 Graph regarding the perception of the landscape quality of the natural environment. Below, the frequency of the answers given at the different dissemination points (elaboration of the author in collaboration with CNRS ESPACE—UMR 7300)

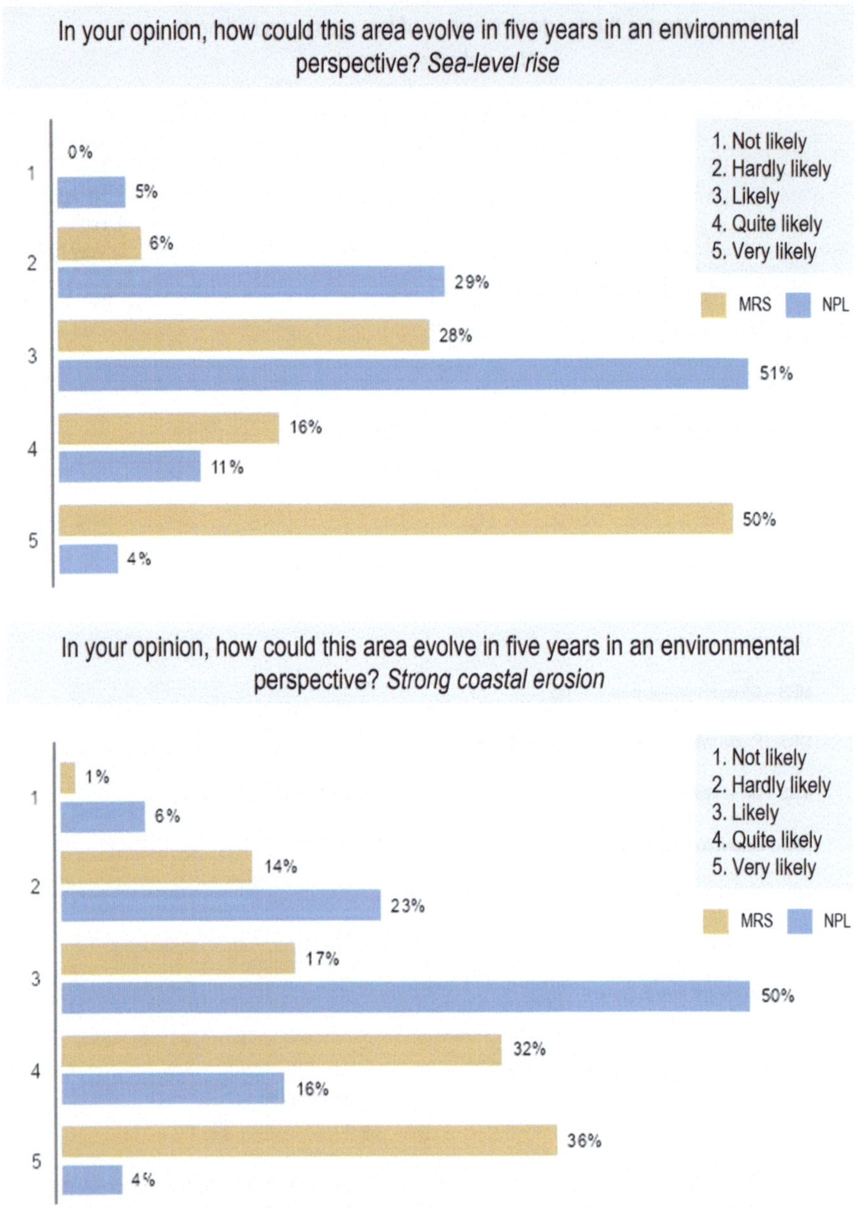

Fig. 14.13 Awareness of risks from the sea diverges between the two urban coastal societies, with Neapolitan citizens displaying greater indifference to environmental criticalities than Marseille users (elaboration of the author in collaboration with CNRS ESPACE—UMR 7300)

years. On the contrary, the Marseille citizens show a more varied range of opinions: 24% assesses the risk as implausible, while 22% considers that an intensification of this kind of environmental hazard is likely to occur during the next five years; 28% assesses the hypothesis as very probable and 25% of the interviewees believes that the Marseille city-sea interface will certainly face an exacerbation of the problem in the short term. Both cities showed widespread disillusionment with the possibility that the next five years will bring an increase in the ecological quality of urban waters and an improvement in marine biodiversity. In the first case, Naples recorded 17% of users completely certain that such eventualities will not occur and 43% showing feeble confidence in such a change in the short term, while only 6% stated with certainty that the urban water quality will improve in the next five years. Similarly, in Marseille, 37% of citizens rates as quite implausible an eventual increase in water quality in the coastal zone, while 34% are moderately confident about it, compared to 15% of the sample who believes that such development is quite likely to happen.

At the same time, about one-fifth of the Neapolitan users thinks that the marine biodiversity will not improve at all in the future and 40% of the sample believes that there are few possibilities for a positive development in this regard; however, 26% is moderately optimistic; the Marseille users are for the most part not very confident (47% of the sample believes that it is unlikely that marine biodiversity will improve within five years), but it is also recorded that 38% of the users thinks that a positive development of the ecological issue may be expected in the next future.

Finally, the questionnaire analysed the social desires of the interviewees regarding the functional evolution of the coast in the next five years, in relation to its socio-recreational functions. Regarding the possibility of reclaiming new walkable spaces from the sea, French citizens show a strong reticence as more than half of them is against to this scenario; 18% would like the urban coast to evolve in this sense and only 6% expresses a strong desire in this regard. Neapolitan users had a similar opinion, with 36% totally against this possibility of urban development and 17% finding it a viable solution; about a quarter of the interviewees expresses an absolutely neutral opinion. With regard to the future possibility of realising floating structures, such as walkways or platforms on the water and solariums, the littoral society in Naples shows considerable interest, with 50% of the users in favour and 16% showing great enthusiasm for the hypothesis; only 7% expresses complete disagreement. For the citizens of Marseille, on the other hand, the possibility of implementing the city-sea interface in the short term with floating structures is not a widely shared scenario: 36% of the users states to be totally unfavourable, while 21% denotes a feeble interest; this potential urban evolution meets a great desire among only 10% of the social sample (Fig. 14.14).

The two studied samples of urban coastal society agree on the hypothesis of increasing public spaces for recreation and socialising in the next five years: data from Marseille show that 54% of users would really like a similar development for the city's coastline, just as in Naples 42% of the interviewees shows great interest in such hypothesis; values about other opinions on this subject tend to be drastically lower in both cities, reaching only 6% in France and 1% in Italy among users that are totally against it. A similar situation occurs with the proposal to design new facilities

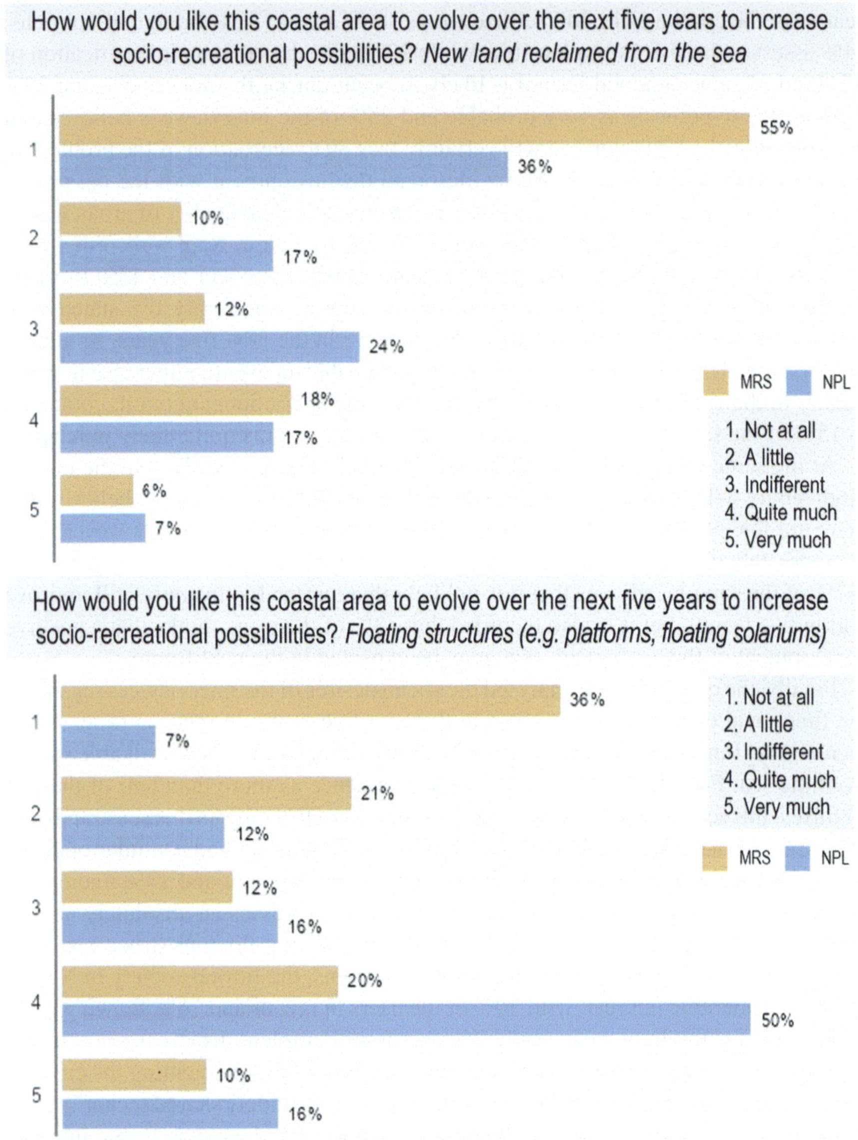

Fig. 14.14 Social desires regarding the short-term coastal evolution. Future land reclamation operations are not particularly appreciated by both coastal societies (elaboration of the author in collaboration with CNRS ESPACE—UMR 7300)

for bathing, with 44% of the users in Marseille and 34% of the users in Naples being very much in favour of this option; here too there are very low values for interviewees showing indifference to this possible development of the city-sea interface, or even aversion. On the contrary, the opinions of Naples and Marseille strongly diverge on the possibility of further extending the areas intended for commercial activities along the urban amphibious in the near future: 58% of the Marseille users completely disagrees, while only 2% seems to be totally in favour of this kind of development; although the Neapolitan users who are absolutely in favour of this hypothesis are only 8%, there is a portion equal to 19% who states to be quite interested in such a coastal transformation, compared to 30% of users who instead remains neutral. With regard to the future realisation of new leisure docks, French citizens appear to be satisfied with the current situation, as 65% of the studied social sample denotes a strong aversion to implementations of existing harbour infrastructures in the short term; Neapolitan users, on the other hand, appear much more interested, with 33% responding in favour of this scenario for the development of the city-sea interface.

14.5 Analysing Coastal Social Demand to Support Policy Makers and Planners

The results of the psycho-social approach showed that in both case studies the city-sea interface is perceived by users as a resource for social and recreational functions and activities. However, there are morphological and functional differences that generate a peculiar way of experiencing and using the urban coasts in the two cities. First of all, it is possible to find in both Naples and Marseille the same desire to be in physical contact with the urban sea: however, although both urban coastal societies state that they consider very relevant the access to the city's waters for social and recreational purposes, in Naples the interviewees tend to frequent the city-sea interface less than the users in Marseille. Moreover, this reflects a paradox that emerged in the social samples of both case studies: accessibility to the sea is often perceived as denied by the presence of constraints to bathing and to the fruition of the shore in general. In particular, the strongest constraint identified by the psycho-social analysis is maritime pollution and poor ecological water quality in the city. In spite of this, it is possible to find a similarity between the urban coasts of the two cities, namely the generally high landscape value attributed to both urbanised and naturalistic areas of the city-sea interface: it should be noted that the environmental quality is high even in points of the littoral that are densely built, as the population of the two cities strongly perceives the presence of the maritime element as a mitigating factor of the anthropic components existing along the urban coastline (Fig. 14.15).

The urban coasts of the two Euro-Mediterranean cities present different degrees of accessibility to the coast, due to obstacles of a physical nature as well, necessarily influencing the frequency of visits to the sea in the city: according to the data collected, these limitations influence the Italian case more than the French one, as Neapolitan

Fig. 14.15 On the left, nautical clubs physically limit accessibility to the sea in the 16th Arrondissement of Marseille. On the right, the urban coast of Bagnoli in Naples is polluted both in water and on land, preventing an equal use of the sea resource (pictures by the author)

users tend to visit the urban coast less often for recreational purposes also due to these constraints. This element leads to a different perception of the activities that are allowed out along the city-sea interface: Marseille users seem to be more aware of the possibility of carrying out recreational activities along the shore such as boating but also fishing, swimming or snorkelling. It is clear, however, that this does not prevent the Neapolitan community from expressing the same socio-recreational needs related to the sea and the coast (Fig. 14.16).

From this it follows that the Marseille and Naples littoral interfaces also differ in their vision of the public coastal space. The Marseille users are particularly reluctant to accept changes to the urban coastline linked to commercial or accommodation or restaurant sector, unlike their Neapolitan counterparts in the survey: in Italy, it seems that the possibility of a transformation oriented towards profit and an increase in tourism can be combined with the recreational offer for local users. Similarly, different opinions have been recorded about the possibility of enhancing infrastructures related to pleasure boating and sailing; the meeting point is found in the common interest for the hypothesis of installing more public facilities that allow for community use of the coastal resource. In both case studies, the future possibility of having more spaces for outdoor sport and bathing activities is welcomed, but users from France and Italy disagree about reclaiming walkable land from the sea or realising floating elements for socio-recreational purposes, indicating a different understanding of the evolution dynamics of coastal areas. It is interesting to note that, although Marseille has experienced several permanent modifications of the coastline in its recent history (think of the transformation of the Prado bathing area), the users of this city do not approve the possibility of further changes to the shoreline profile, unlike the citizens interviewed in Naples (Fig. 14.16).

From the environmental point of view, the users interviewed in Naples and Marseille share the same doubts about the future environmental improvement of the city-sea interface: both of the analysed social samples show a certain reticence in believing that in the short term the environmental quality of the urban coastline can

Fig. 14.16 Despite the evident constraints, the social desire to enjoy the coast and the sea of the city-sea interface still appears strong. In the picture, leisure docks in L'Estaque, Marseille (picture by the author)

be increased, just as they believe that marine biodiversity is unlikely to be enriched. However, they have a different understanding of the risks linked to climate change and the threats coming from the sea that undermine the development of the urban amphibious: in particular, the answers provided in Naples show a general lack of awareness of the effects of phenomena such as coastal erosion or the progressive sea-level rise, or at most a general indifference; on the contrary, in Marseille a more solid attention to contemporary environmental criticalities emerged.

The use of the psycho-social approach is thus useful in assessing the social representation of urban coastal society: its needs can be empirically studied through a socially variegated sample, analysing the degree of connection between the city and the sea. This type of investigation allowed the comparison of the coastal interfaces of two large cities within the Euro-Mediterranean basin: as shown above, the flexibility of the approach enables the structure of the analysis method to be specifically shaped for benchmarking activities between other coastal urban realities. Although the two case studies present similarities, it is evident that the social demand is different; however, it emerges a marked propensity to implement public interface space for sociality that should be more integrated within the management and planning of coastal areas. In fact, in addition to the comparison of the case studies, which provides relevant analytical insights, the data collected can be included within the design and urban planning processes to diversify the functional offer of the city-sea interface from a community perspective, through social listening and interpretation of the social demand. In this way, the analytical tool becomes an operational basis that can concretely support the decision-making dynamics and the planning of the coastal public space in an adaptive manner.

Chapter 15
Conclusions and Research Perspectives

Abstract The main objective of this volume is to demonstrate that the concept of city-sea interface could be intended as a bridging point between coastal planning and maritime management, through a planning downscaling process that could allow the interaction of coastal community demands with ecological-environmental issues within the local and sub-local decision-making frameworks. The innovation of the proposed work lies in the possibility of adaptively supporting the realisation of quality coastal public spaces, in order to enable a real urban ecological transition, through the flexibility of the urban amphibious and the methodological framework that has been tailored to it.

Keywords Socio-environmental well-being · Downscaling · Coastal spatial planning · Proximity public spaces · Governance support · Local-scale design

15.1 Final Results: Overcoming Large-Scale Coastal Planning for Local Socio-Environmental Well-Being

Contemporary coastal cities are characterised by the increasingly complex issues that mainly influence the development of the most densely urbanised and inhabited ones. The book has investigated criticalities related to the use of social spaces within articulated urban contexts such as the city-sea interface, proposing a systemic interpretation of its peculiar characters and deepening technical-operational tools to mediate common dynamics within the innovative maritime spatial planning and the more traditional coastal land planning. Specifically, the Euro-Mediterranean context is particularly interesting due to the presence of a very high demographic concentration and crucial socio-cultural, landscape and environmental values. The originality of this volume lies precisely within this branch of the urban planning discipline, which is currently at the centre of the scientific debate on the sustainable evolution of coastlines. The proposed theoretical declination revolves around the hinge of social demand for public space that can be equally enjoyed by the coastal community, in accordance with the current ecological transition, which imposes on the one hand

I. Pistone, *Urban Coasts in Socio-ecological Transition*,
https://doi.org/10.1007/978-3-031-70783-4_15

the use of nature-based solutions that minimise impacts on coastal areas, and on the other an adequate level of attention to plan adaptation strategies in relationship with the constantly worsening climate context. The main outcomes of the present work are therefore described, in the context of the aforementioned critical issues.

1. *The city-sea interface concept can be read as a synthesis of traditional coastal spatial planning and the more innovative maritime spatial planning, integrating land-sea interactions within management and planning processes.*

The city-sea interface is characterised by the presence of complex social and ecological relationships that influence the local communities on a daily basis and that have determined its peculiar urbanisation. The effects of these phenomena are clearly reflected on the liveability of the urban coastline, but also on the environmental system, composed equally of the terrestrial and marine sides of the shore. As a matter of fact, the city-sea interface, in the sense of urban amphibious, hosts the community dynamics underlying urban coastal society and at the same time the stratification of coastal plans that are not always coherent and homogeneous. The objective of this theoretical concept is therefore to bring out the littoral flexibility, exploiting its potential from a planning perspective, but at the same time to define a methodological order for the management of the three edges composing the city-sea interface, described in Chap. 2 of this text: the maritime sphere of the sea edge should therefore not be considered disconnected from coastal land use planning, just as the maritime city edge constitutes the functional link between the inner areas of the extended city edge and the water system, from a socio-cultural, touristic, ecological and environmental point of view. In this context, open spaces, urban green areas, proximity public spaces, as well as maritime areas near the shore with cultural and recreational functions or potential represent opportunities for ecological planning, paving the way for the social and sustainable design of coastal areas in highly urbanised areas, transcending the ephemeral spatial limits between land and water: indeed, the context of the urban littoral is becoming increasingly inclusive, encouraging planning actions that provide users with a physical-visual connection to coastal-marine environments through ecologically dynamic and multifunctional spaces (Beatley 2018). It is also true that in coastal spaces, both terrestrial and maritime, many different uses coexist, often concentrated in a relatively small area, including littoral tourism, cultural functions related to heritage, protected areas, fishing, aquaculture, but also mining and energy collection, military uses and transport of goods and passengers. The importance of a careful but at the same time flexible regulation that allows the fruition of resources and the protection of the environment becomes evident (Barbanti et al. 2015). In this sense, coastal management can benefit from the adaptive character of the urban amphibious in order to efficiently orient land-sea interactions in the context of maritime and coastal spatial plans. However, as highlighted in Chap. 7, little steps have been taken in most of the Euro-Mediterranean countries considered in this book. France has shown a greater propensity for this type of innovative planning of the maritime and littoral space, through the *Stratégie Nationale Pour la Mer et le Littoral*: this instrument emphasises the issues of ecological transition and water quality, but also the socio-tourist organisation of the shore and the sustainability of sea uses. France can thus be a model for the other analysed states that are moving, at different stages, towards an organic and cohesive management of their coastal maritime domains, approaching international cooperation processes.

2. *At present, maritime spatial planning suffers from a lack of implementation at the local scale, making downscaling necessary to bring the issue back into the social and community perspective of the urban amphibious.*

The public spaces within the city-sea interface are linked to the interests of different categories of stakeholders, with regard to the land and the water side. However, one of the critical nodes of the integration between land-based urbanism and modern maritime spatial planning concerns the difficulty for the users of the urban coastal society to perceive tangible effects on the littoral territory, limiting the interest and motivation of the community that should instead be concretely involved in such processes. In fact, strategic orientations often appear to be cast from above and tend to be defined at the national or regional level rather than at the local scale (Zaucha and Gee 2019). This clearly implies different levels of understanding of policies drafted to incorporate land-sea interactions within conventional planning instruments, especially in the present context where traditional land-use planning and maritime spatial management may sharply diverge, undermining the overall functional cohesion of the city-sea interface: it is therefore necessary to organise in a conscious manner different requirements at various scales within the articulated coastal planning system (Morf et al. 2022). In order to implement maritime-coastal spatial plans, downscaling processes that bring the attention of administrations and planners back to the local interests of communities are crucial. In order to ensure an integrated development of the urban amphibious, it is necessary to produce development scenarios of the desired land and maritime activities, with the aim of directing functional growth towards socio-ecological sustainability, analysing the peculiar environmental characteristics and expected impacts, mediating existing conflicts and enhancing synergies between uses and social groups of the urban coastal society. In this sense, the tourist-cultural sphere can act as an engine for economic development without compromising the natural resources and coastal commons offered by the city-sea interface. The comparison between similar cases can provide important indications on how to operate in favour of coastal defence, positive delocalisation of coastal and sea uses and environmental protection: the socio-recreational functions can be coherent with low-impact technologies, such as sustainable energy production (such as thalasso-thermal energy), but also with activities like fishing or non-intensive fish or mussel farming. All these elements are consistent with the principles of blue growth on a local scale, provided that these functions are logically organised in relation to the most intensive uses of the urban littoral, such as port activity, and that targeted measures are identified to safeguard the coastal environment (Barbanti and Perini 2018). From an operational point of view, the downscaling process involves the identification of data to describe at the local scale the urban amphibious: the integration of existing datasets with specific elements related to the socio-recreational use and geomorphological coastal characters can define a relevant analytical starting point for future plans and projects.

3. *Bringing urban amphibious planning back to the human and local scale favours the development of high-quality proximity public spaces, responding to the social demand for urban well-being and to the ongoing ecological transition.*

The emerging social demand for proximity public space goes hand in hand with the gradual scale reduction required by coastal planning: from a community and recreational point of view, it has emerged how the ecological areas of the city-sea interface can satisfy the specific needs of urban coastal society in terms of sociality and urban well-being. The concept of proximity, and therefore of small scale, within the interface project can be a valid element in mending the flaws inherent in urban coastal society, linked to unequal accessibility to the sea and its resources. The possibility to access, within a short distance, to areas available for human interaction and enjoyment of coastal naturalness in the context of large cities is part of the 15-minute city theme (Moreno 2020) and allows to reduce the need to move outside the main urban cores for the fruition of high-quality littoral assets, in the perspective of the ongoing ecological transition; at the same time, it protects the right to relational proximity, that is, the character that public spaces must necessarily offer in terms of sociality with other individuals, strengthening the shared identity of the community (Manzini 2021). Such reasoning gains even more value in relation to the post-pandemic period in which we live and which has stimulated reflection on the distributive justice of places, as well as their wholesomeness: the impact of this unusual criticality has transformed even open spaces into a kind of gated communities, urban islands where sociality can be at risk but can also constitute a risk (Kleilein and Meyer 2021). Health effects are thus one of the cornerstones that the ecological planning of the city-sea interface must promote. In this sense, a responsible design of urban shores can provide efficient solutions, calibrating the amount of access to urban water, fairly sizing social spaces and mending the unstable connections between green zones, dense areas and closely related coastal spaces in order to consolidate them within the broader framework of green–blue infrastructure within the field of ecological planning. As it has been described, it is precisely the open and less urbanised spaces that are more prone to transformations, a potentiality that can easily turn into a weakness in relation to both the interests at the base of coastal functional competition and the growing environmental criticality. On the basis of the reasoning of Chap. 3, it is necessary to emphasise an integrated approach that recognises the existing interdependencies between environmental, social and technical factors, recognising at a multi-scalar level the potential vulnerabilities of the different areas of the coastal interface. Adaptive environmental planning requires to manage the coastal resource carefully considering existing social and institutional arrangements and user awareness in relation to the new environmental condition (Lloyd et al. 2013). The recomposition of maritime-coastal ecosystems through the active involvement of urban coastal society can thus lead to a more equal and rational organisation of public spaces and the related coastal commons, limiting the effects of environmental risks, also by exploiting technologically innovative solutions. This clearly has an important influence on the inclusion of the dynamics of land-sea interactions within the ecosystem approach to maritime-coastal spatial planning, introducing such criteria into the governance processes that should facilitate their implementation (Bąkowski and Nyka 2022). The results of the work set out in Chaps. 5 and 7 show, however, that France once again represents the most advanced model in the field of urban coastal planning and management for social and recreational purposes, in the Euro-Mediterranean context, as it transposes EU directives and at the same time limits negative changes on coastal land as much as possible, establishing specific regulatory systems aimed at liberalising public coastal space and improving the supply of socio-recreational services. The Italian case, on the other hand, appears particularly complex, since the coastal management matter still seems vague and is burdened by Community sanctions.

4. *The proposed methodological framework, applied to the city-sea interfaces of Naples and Marseille, offers indications for planning downscaling, interpreting the needs of urban coastal society in the light of the existing normative system and the spatial-functional features of the urban amphibious.*

From an operational point of view, the definition of the methodological framework, described in Chap. 10, allows to simplify benchmarking operations between similar urban coasts belonging to the same geographic-functional context. Within the Euro-Mediterranean context for which it was conceived, this approach allows for the analysis of the regulatory and governance processes currently active along the reference city-sea interface, in order to implement the active participation of urban coastal society within existing coastal planning, consequently improving the social aspects of urban amphibious. The analysis of the relationships between land and sea, conceptualised by the spatial components of the city-sea interface, can be carried out through the geo-spatial and psycho-social components of the framework, which brings together different types of approaches to verify on the one hand the existing social demand in terms of areas usable by the local community, and on the other hand the actual functional, environmental and morphological availability of the urban coasts to host present and future functions of urban sea resources and littoral places. In Chaps. 12, 13 and 14 of this text, the approach was applied, on an experimental basis, to compare the normative, functional, morphological and social characteristics of the coastal interfaces of Marseille and Naples, understood as large coastal cities whose actual similarities and differences in the management and community perception of the coastal public space have been analysed. From the normative point of view, the volume showed that Marseille could be considered as a model of Euro-Mediterranean city that has been able to renew itself in the last decades: it presents in fact a range of directives and plans aimed at the qualitative improvement of the city-sea interface, taking care of its different aspects (social, recreational, environmental, cultural). Naples, on the other hand, tends to be more static, with the presence of numerous projects that have remained on paper; moreover, it has to be highlighted the unusual administrative overlap between the municipal authority and the port authority with regard to the management of the maritime domain: if this leads to a general municipal lack of interest in the development of certain coastal spaces intended for the urban coastal society, it is nonetheless evident that the Port Authority itself outlines, at least at a strategic level, guidelines for the reconnection of land and sea from a social point of view. In both cities, however, coastal transformations mainly result from top-down planning, leading to an uneven involvement in social and recreational themes of urban amphibious. From a geo-spatial perspective, it emerged that in part the different management of the city-sea interfaces of the two cities depends on the morphological conformation of their urban coastlines, specifically of the maritime city edge, the middle area of the city-sea interface, considered as the connecting element between land and urban sea where the main socio-environmental coastal interactions are concentrated. Marseille presents a particularly reduced maritime city edge in terms of size, as its transit road (the first road infrastructure parallel to the coast that interrupts the ecological and functional continuity within the littoral interface) is generally very close to the coastline; moreover, the central portion of the study area is occupied by the massive Grand Port Maritime de Marseille. In spite of this, the French city demonstrates a certain ability to organise the reduced spaces for bathing and outdoor sports activities. In Naples, the maritime city edge has a larger surface, especially in the Posillipo area, but accessibility to the coast is often blocked by obstacles of various kinds (linked to the privatisation of coastal spaces or to governance dynamics); besides, it is not uncommon that the spaces destined for social use are lacking in terms of public services and equipment. Both in Italy and in France, however, there is a limited percentage of public spaces destined for urban coastal society: Naples, although less evolved from a planning point of view, nevertheless shows a wider spatial potential than Marseilles, as well as greater coastal naturalness. Finally, from a psycho-social point of view, the survey showed that in both cities the city-sea interfaces possess great value for sociality and recreation in the eyes of the users involved in the fieldwork. In any case, Neapolitans are less likely to frequent the urban coastline than their counterparts in Marseille, even though in both case studies there is a shared perception of inaccessibility to the sea, mainly due to poor water quality and environmental pollution: the data show, however, that the landscape quality of the built and natural environment of the urban amphibious is considered of great value by the French and Italian samples. On the basis of the information collected in different areas of the urban coasts of Marseille and Naples, it is possible to state that the methodological framework can also be considered as a possible solution to the abovementioned downscaling problem inherent in the implementation of coastal-maritime spatial planning. Indeed, maritime spatial planning still has to deal with a very large scale of application and needs to downscale at the local level in order to actively involve stakeholders and efficiently organise community uses by incorporating land-sea interactions into the drafting of plans. The innovation of the methodological proposal lies in the possibility of reading the city-sea interface data as an analytical basis for understanding socio-ecological demand from a planning perspective, serving as an operational support for downscaling future coastal projects, plans and policies. In fact, the analysis has shown an interesting social and functional background regarding Marseille and Naples, highlighting merits and demerits of the current regulatory framework and how the urban coastal society perceives the actual organisation of uses, the coastal landscape and the environmental conditions: these elements can be read as an information structure at the basis of the future planning development in a socio-ecological perspective for the two cities and for the city-sea interface.

15.2 Future Perspectives. Indications for Implementing the Methodological Framework to Ground Coastal-Maritime Planning and Management

Within the research process, the methodological framework was designed to study and understand the emerging needs for public spaces and places of sociality, in relation to the morphological and functional characteristics of the city-sea interface and the regulatory frameworks in force, taking into consideration the increasingly topical climate and environmental issues: this is also linked to the particular pandemic timeframe during which part of the work was carried out, a period that highlighted the importance of allocating a fair use of the places in the context of the urban amphibious. From a planning point of view, the volume offers several possible hooks to further develop the presented themes and broaden its scope, thanks, moreover, to its strongly multidisciplinary context.

In fact, the described methodological framework has a high degree of flexibility and adaptability: this allows the field of research to be extended to further functional areas of the city-sea interface. As described in the Part I of this book, the city-sea interface encompasses interactions between the environmental system and the urban-anthropic system: these can diversify into numerous branches of urban coastal uses. Functions such as non-intensive fishing, transport by land or sea, aquaculture and mussel farming can be included in the list of fields that can be analysed with the proposed framework; in a perspective of sustainable and ecological development, these uses of the coast and the sea could provide an important benefit to the wider coastal system. From the psycho-social point of view, the framework can be calibrated to explore the needs and perceptions of different types of users, in different geographical contexts: actually, the survey model has a high degree of adaptability, allowing to apply the method to different coastal realities, even smaller ones. From a geo-spatial point of view, it is possible to extend the layers composing the spatial data model in order to create shared databases on the main coastal urban centres within the Euro-Mediterranean basin. It is also conceivable to include environmental aspects of the city-sea interface in a more in-depth manner, such as water quality or climate risk elements. In relation to the latter parameter, the analysis of the direct or indirect impacts on the public littoral space, caused by the growing environmental crisis, may benefit from an information system that relates the different interdependent systems composing the city-sea interface, helping urban planners and policy-makers in estimating the best planning solutions and responses to face direct or indirect damage to the natural, social or even economic spheres of the urban amphibious: the geo-spatial approach of the present volume could thus evolve into a proactive support to ecological planning (Raven et al. 2018).

Beyond its use as a benchmarking tool (which in any case constitutes an important output of this volume), the methodological framework can be read as an attempt to provide a solution to the issue of downscaling for maritime-coastal spatial planning. The attention paid to the interpretation of the demand for spaces dedicated to sociality and the landscape-environmental perception of the city-sea interface and its criticalities can in fact offer an important analytical contribution for the drafting of future plans aimed at the integrated and ecological implementation of urban coasts. The normative analysis of the existing legal framework, the geo-spatial approach and the psycho-social approach can in fact be combined in a complementary way in order to elaborate a unified operational method that can elaborate spatial-functional data with social perceptions, integrating this knowledge with the actual policies and plans. The integration of the these approaches could allow to identify neuralgic hotspots along the city-sea interface, descending with even greater precision to the urban scale, spatially localising the perceptions and needs recorded by field surveys and cross-referencing them with spatial data, verifying the actual lack of services, interpreting the characteristics of the urban littoral in order to verify the feasibility of interventions on public spaces and maritime domain, in accordance with current regulations and any policy and plan in progress. According to the principles of environmental psychology and landscape research, in fact, the transformation of urban land is related to the attitude that users show towards a given change, a factor that strongly influences the success of a specific intervention, but can also direct future planning choices: this is particularly relevant for the blue-green space project, as the combination of quantitative data, obtained from geo-spatial mapping and the elaboration of questionnaires disseminated among citizens, can stimulate the qualitative analysis for the re-functionalisation of urban open areas, taking into account the multidimensional variables that each user attributes to them (Stessens et al. 2020). Clearly, these are highly collaborative techniques, which link the various branches of the approach presented in this book: the characteristics of the existing coastal public spaces can condition the formulation of the psycho-social survey, just as the wishes expressed by the users (and the normative indications) can determine how the data should be collected and how the layers of the spatial data model have to be defined (Balram and Dragićević 2005).

This methodological framework could also be related to the use of other small-scale digital tools and technologies to concretely and operationally implement downscaling processes aimed at maritime-coastal spatial management, from the planning of spaces to the definition of the specific functional elements. The aim is that the proposed approach can become a support basis for local planning, within the framework of already existing coastal tools: referring to the case studies of the volume and, in particular, to Chap. 12 of the text, in Naples, the framework could stimulate the application of the *Piano Attuativo di Utilizzazione delle Aree del Demanio marittimo* (PAD), providing useful information for the allocation of concessions and the definition of spaces to be designed for socio-ecological functions, in accordance with the desires of the urban coastal society; in Marseille, it is possible to hypothesise an integration with the *Plan Plages et Littoral*, which could become in this way more efficient in the identification of critical points of the coast and in the localisation of potential new accesses to the maritime resources and services related to the landscape, environmental and socio-recreational enhancement of the city-sea interface.

Therefore, with the objective of implementing the proposed methodological model, in view of the conceptual outcomes that emerged from this work, the further improvement of an analytical and design approach for the city-sea interface is identified as the main axis of future development. Since the methodological framework offers the possibility of reading the urban amphibious from a legal, environmental, planning and social perspective, it could be useful as a base for sustainable planning of the coast and the urban sea that respects the local scale of littoral uses proper to the urban coastal society, which has an increasingly central role within these complex management processes.

References

Bąkowski T, Nyka M (2022) Land-Sea Interactions in Realisation of Ecosystem Approach in the Marine Spatial Planning in the Baltic Sea Region – Polish Perspective. Review of European and Comparative Law 51(4): 209–236

Balram S, Dragićević S (2005) Attitudes toward urban green spaces: integrating questionnaire survey and collaborative GIS techniques to improve attitude measurements. Landscape and Urban Planning 71(2): 147–162

Barbanti A, Perini L (eds) (2018) Fra la terra e il mare. Analisi e proposte per la pianificazione dello Spazio Marittimo in Emilia-Romagna. Zenodo

Barbanti A, Campostrini P, Musco F, Sarretta A, Gissi E (eds) (2015) Conclusioni e raccomandazioni del Progetto ADRIPLAN. Un manuale breve per la Pianificazione dello Spazio Marittimo nella Regione Adriatico-Ionica. CNR-ISMAR, Venice

Beatley T (2018) Blue Biophilic Cities. Nature and Resilience Along the Urban Coast. Palgrave Macmillan, London

Kleilein D, Meyer F (eds) (2021) Post-pandemic Urbanism. JOVIS, Berlin

Lloyd MG, Peel D, Duck RW (2013) Towards a social–ecological resilience framework for coastal planning. Land Use Policy 30(1): 925–933

Manzini E (2021) Abitare la prossimità. Idee per la città dei 15 minuti. Egea, Milan

Moreno C (2020) Droit de cité: De la “ville-monde” à la “ville du quart d’heure”. Editions de l’Observatoire, Paris

Morf A, Moodie J, Cedergren E, Eliasen SO, Gee K, Kull M, Mahadeo S, Husa S, Vološina M (2022) Challenges and Enablers to Integrate Land-Sea-Interactions in Cross-Border Marine and Coastal Planning: Experiences from the Pan Baltic Scope Collaboration. Planning Practice & Research 37(3): 333–354

Raven J, Stone B, Mills G, Towers J, Katzschner L, Leone M, Gaborit P, Georgescu M, Hariri M (2018) Urban planning and design. In Rosenzweig C, Solecki W, Romero-Lankao P, Mehrotra S, Dhakal S, Ali Ibrahim S (eds) Climate Change and Cities: Second Assessment Report of the Urban Climate Change Research Network. Cambridge University Press, New York, p. 139–172

Stessens P, Canters F, Huysmans M, Khan AZ (2020) Urban green space qualities: An integrated approach towards GIS-based assessment reflecting user perception. Land Use Policy 91

Zaucha J, Gee K (eds) (2019) Maritime Spatial Planning. Past, present, future. Palgrave Macmillan, Cham